Lecture Notes in Operations Research is an interdisciplinary book series which provides a platform for the cutting-edge research and developments in both operations research and operations management field. The purview of this series is global, encompassing all nations and areas of the world.

It comprises for instance, mathematical optimization, mathematical modeling, statistical analysis, queueing theory and other stochastic-process models, Markov decision processes, econometric methods, data envelopment analysis, decision analysis, supply chain management, transportation logistics, process design, operations strategy, facilities planning, production planning and inventory control.

LNOR publishes edited conference proceedings, contributed volumes that present firsthand information on the latest research results and pioneering innovations as well as new perspectives on classical fields. The target audience of LNOR consists of students, researchers as well as industry professionals.

Constantin Zopounidis · Liadaki Angeliki ·
Eskantar Marianna
Editors

Operational Research Methods in Business, Finance and Economics

Proceedings of the 31st European Conference on Operational Research, Athens, Greece, July 11–14, 2021

Springer

Editors
Constantin Zopounidis
School of Production Engineering
and Management
Technical University of Crete
Chania, Greece

Audencia Business School
Nantes, France

Liadaki Angeliki
Financial Engineering Laboratory
School of Production Engineering
and Management
Technical University of Crete
Chania, Greece

Eskantar Marianna
Financial Engineering Laboratory
School of Production Engineering
and Management
Technical University of Crete
Chania, Greece

ISSN 2731-040X ISSN 2731-0418 (electronic)
Lecture Notes in Operations Research
ISBN 978-3-031-31240-3 ISBN 978-3-031-31241-0 (eBook)
https://doi.org/10.1007/978-3-031-31241-0

This Springer imprint is published by the registered company Springer Nature Switzerland AG
The registered company address is: Gewerbestrasse 11, 6330 Cham, Switzerland

Preface

Operational Research (OR), as a multidisciplinary analytical field, contributes in many ways in the design, dissemination, and management of several new technological innovation in business, finance, and economics. Among others, these include areas such as big data analytics, green technologies, socially responsible investments, supply chain networks, economic efficiency, warehouse management systems, banking management, risk assessment, and energy systems.

This volume is presenting advances on the contributions of OR approaches in technology-driven areas in business, finance, and economics, covering both theoretical/methodological developments and real-world case studies. This volume includes revised and substantially extended versions of selected papers presented at the 31st European Conference on Operational Research (EURO Athens, 2021 Greece).

This book will incorporate the interest of students, academics, and professionals of different industries and organizations.

An initial table of contents is as follows.

The first chapter "Stochastic Differential Equations in L^p-Spaces" provides more specific results on stochastic integration and stochastic differential equations on Lp-spaces, while the results are extensions of equivalent results on Brownian stochastic differential calculus. Their applications are related to financial markets' structure.

The second chapter "A Machine Learning Approach to Entrepreneurial Finance Modelling" is an in-depth examination outlining methods and approaches for application of segmented modelling in entrepreneurial finance, as well as how they can be applied using existing data for purposes to examine selection, valuation, and survivability.

The third chapter "Green Versus Non-green Banks: A Differences-In-Differences CAMEL-Based Approach" examines whether there are any discernible performance differences between green and non-green banks. The variables of interest are fundamental CAMEL factors. By employing panel data techniques, authors investigate whether there are statistically significant differences between the two groups. Among the main results that authors have concluded is that green banks, whether global or not, generally do not differ from their non-green counterparts in terms of CAMEL ratios before and after the financial crisis. They also find that the crisis has equally affected the ratios of both bank types, either positively or negatively: (a) a positive effect on capital adequacy, asset quality, and management quality and (b) a negative effect on earnings ability and on liquidity.

The fourth chapter "Measuring Corporate Gender Diversity and Inclusion with UW-TOPSIS and Linguistic Intervals" proposes fuzz adequacy indicators to measure the degree of gender diversity in firms. The construction of these indicators is based on the multiple criteria decision-making method which is an extension of the Unweighted Technique for Order of Preference by Similarity to Ideal Solution (UW-TOPSIS). The proposed method allows the ranking of the decision alternatives without a priori determination of a precise weighting scheme. The main contribution of the paper is the use of linguistic labels transformed into linguistic intervals incorporated into the UW-TOPSIS algorithm to rank a set of decision alternatives. With this method, the relative proximity to the positive ideal solution is optimized for each firm based on the possible linguistic intervals expressing the criteria weights. As a result, authors obtain a relative proximity interval informing the decision-maker about the worst and best possible positions of each firm in the ranking. This could provide a useful information in terms of improvement opportunities for the firms and allows the decision-maker to express certain preferences regarding the decision criteria using linguistic terms.

The fifth chapter "A Multicriteria Analysis Approach to Tourists' Satisfaction with Local Food Consumption" studies tourists' satisfaction with local food consumption. Greece is selected as a case study because of the importance of its culinary tradition. The analysis is based on an extension of the multicriteria analysis MUSA methodology. The results of this analysis allow several conclusions about tourists' satisfaction from local food consumption. Most importantly, this study confirms the importance of sensory traits, like taste, while authenticity and connection to Greek culture are also important for customer satisfaction.

The sixth chapter "Ecotourism as a Tool for Regional Development in the Area of Prespa National Forest Park" aims to study the area of the Prespes National Park as an ecotourism destination and to promote the provided alternative activities on the Internet. The websites that promote the tourism enterprises of the area are analyzed and classified using the method of multicriteria analysis PROMETHEE II. Finally, suggestions for the optimization of the existing websites are presented in order to enhance the online promotion of the area of Prespa National Forest Park.

The seventh chapter "Evaluating the Importance of ESG Criteria: A Multicriteria Approach" focuses on the question that arises whether all companies regardless of the sector or country they belong should publish the same ESG indicators. ESG criteria are considered an important topic that all businesses worldwide should consider. Businesses should be able to publish this kind of information so that one can consider its viability in matters concerning the three pillars. In this paper, the ESG criteria were examined over time, dividing the period 2007–2019 into three equal periods, for each pillar separately, for 39 countries worldwide (developed and developing) and 16 industries. The results indicate that the social dimension is the most important. The environmental pillar was also found to be of high importance, thus indicating that climate change is now affecting corporate decisions and investors. As far as the governance pillar is concerned, it appears to be of lower importance compared to the other two pillars.

The eighth chapter "Lufthansa Airlines. The Microeconomic and Macroeconomic Environment of the Company and the Industry in 2020 and Its Readiness Against Crisis" provides an analysis of the impact of the microeconomic and macroeconomic environment of Lufthansa Airlines and sets out the performance of the organization in relation to its main competitors for a specific time period. In order to explore the company's market exposures and its cost vulnerabilities, a dedicated analysis of the top ten airlines in the world is taking place.

Chania, Greece/Nantes, France — Prof. Constantin Zopounidis
kzopounidis@tuc.gr

Chania, Greece — Liadaki Angeliki, M.Sc.
aliadaki1@tuc.gr

Chania, Greece — Eskantar Marianna, M.Sc.
meskantar@tuc.gr

Contents

Stochastic Differential Equations in L^p-Spaces

Christos Floros, Konstantinos Gkillas, and Christos Kountzakis

Abstract In the present paper, we do provide more specific results on stochastic integration and stochastic differential equations on L^p -spaces. We also may obtain the existence and uniqueness of strong solutions to these stochastic differential equations. These results are actually extensions of equivalent results on Brownian stochastic differential calculus. Their applications are related to financial markets' structure, since this structure includes both investment and claim payments on insurance contracts.

Keywords Partially ordered linear spaces · Stochastic integration · Stochastic differential equations · Applications in finance and actuarial science

MSC (2020) classification 46A40 · 44B40 · 60H05 · 60H10 · 91G15

1 Motivation and Further Research

The usual approach in stochastic differential equations is the one appearing under the assumption that both integrated and integrator stochastic processes have some special properties. Such properties' examples are that the integrator stochastic process is a semimartingale, or a local martingale. The references at the end of the paper indicate this standard framework. These assumptions are eliminated in the present paper. The assumptions posed in the present paper are related to the moments of the sums, which formulate the associated stochastic integral is related to the lattice structure of the L^p-spaces. Further research may be related to the analogue of Brownian weak solutions for some stochastic differential equation. Another direction of research is the order boundedness properties of the solutions to these stochastic dif-

C. Floros (✉) · K. Gkillas
Accounting and Finance, Hellenic Mediterranean University, Estavromenos, 71410 Heraklion, Greece
e-mail: cfloros@hmu.gr

C. Kountzakis
University of the Aegean, Mitilini, Greece

C. Zopounidis et al. (eds.), *Operational Research Methods in Business, Finance and Economics*, Lecture Notes in Operations Research,
https://doi.org/10.1007/978-3-031-31241-0_1

ferential equations. Notions related to order boundedness and order completeness are extensively explained in Aliprantis and Border (2006) end especially in Chap. 8. Wong and Zakai (1965) is actually a seminal one in this topic. We do provide a generalization of this paper's framework, since in the above paper, the definition of the stochastic integral is achieved in terms of L^2-convergence; see p. 215 of Wong and Zakai (1965). In the present paper, the convergence is not necessarily the L^2 one, but the L^1-convergence. In such a way of definition, the Riemann–Riesz integral includes stochastic processes with a quite heavy-tail behavior of their random variables. Moreover, in Wong and Zakai (1965) the integrand used is continuous with respect to the Lebesgue measure; see p. 214. Such an assumption may be considered to be restrictive for the same reason. The well-known Itô isometry implies that the behavior of the increments of a stochastic process is similar to the one of Brownian motion, which is not always true. The calculative Lemma (E) of the present paper is a generalization of the infinitesimal operator for Itô processes; see p. 217 of Wong and Zakai (1965). The stochastic differential calculus rules in the present paper are quite similar to the ones appearing in the present paper.

2 Introduction

The establishment of the **Riemann–Riesz** stochastic integral in order complete vector lattices was presented in Floros et al. (2021). Essential notions about vector lattices and order complete vector lattices appear in the Appendix. For calculations' scopes arising below, we assume that this order complete vector lattices are some L^p space, such that $1 \leq p < +\infty$. In the paper, we did defined the Riemann–Riesz integral on L^1-valued stochastic processes. If $1 < p < \infty$, then the Hölder Inequality implies that $p = \frac{p}{p_1} + \frac{p}{q_1}$, where p_1, q_1 are conjugate exponent indices, namely $+\infty > p_1, q_1 > 1$ and $\frac{1}{p_1} + \frac{1}{q_1} = 1$. The applications of the Riemann–Riesz integral in actuarial and financial mathematics include either value or discounted value processes, or surplus or deficit process, since financial institutions do include insurance structure. The advantage of stochastic differential equations with respect to the Riemann–Riesz integral is that they do provide a unified way for the description of the associated stochastic processes. The jumps' component is not separated from the stochastic integration part in the Riemann–Riesz integration. This fact simplifies stochastic integration. The most important works on stochastic integration do appear in the References' part. We consider some probability space $(\Omega, \mathcal{F}, \mathbb{P})$ and some filtration $(\mathcal{F}_t)_{t\in[0,T]}$, where $[0, T]$, $T > 0$ denotes the time horizon of the corresponding stochastic processes. The cardinal number of Ω is assumed to be the one of real numbers and $\mathcal{F}_T = \mathcal{F}$. $L^p = L^p(\Omega, \mathcal{F}, \mathbb{P})$, and we also assume that any random variable Y_t is $\mathcal{F}_t$ measurable, where $Y = (Y_t)_{t\in[0,T]}$ denotes some real-valued stochastic process.

3 Stochastic Calculus Under Riemann–Riesz Integration

We suppose that $X_{t+h} - X_t \in L^p$ for any $t \in [0, T]$ and $h > 0$, such that $t + h \in [0, T]$.

Definition 3.1 $\mathrm{d}t \cdot \mathrm{d}t = lim_{h\to 0^+} h \cdot h = 0$.

Definition 3.2

$$\mathrm{d}t \cdot \mathrm{d}X_t = \mathrm{d}X_t \cdot \mathrm{d}t = lim_{h\to 0^+}\mathbb{E}(X_{t+h} - X_t)h = lim_{h\to 0^+} h\mathbb{E}(X_{t+h} - X_t) = 0.$$

Definition 3.3

$$s_t := \mathrm{d}X_t \cdot \mathrm{d}X_t = \limsup_{h\to 0^+} \mathbb{E}(X_{t+h} - X_t)^2.$$

We notice that the calculation results arising above are **real numbers**, under the assumption that the above moments do exist. This is true if $p \geq 2$.

The above calculation rules may arise, under the following similar form :

We suppose that $X_{t+h} - X_t \in L^p$ for any $t \in [0, T]$ and $h > 0$, such that $t + h \in [0, T]$.

Definition 3.4 $\mathrm{d}t \cdot \mathrm{d}t = \lim_{h\to 0^+} h^2\mathbf{1}$ where 0 is the zero element of L^p.

By $\mathbf{1}$, we denote the element of L^p, which is equal to 1, $\mathbb{P}$ a.e.

Definition 3.5

$$\mathrm{d}t \cdot \mathrm{d}X_t = \mathrm{d}X_t \cdot \mathrm{d}t = lim_{h\to 0^+} h((X_{t+h} - X_t) = lim_{h\to 0^+} h(X_{t+h} - X_t) = 0,$$

where 0 is the zero element of L^p.

The above rule arises from the structure of L^p as a linear space.

Definition 3.6

$$s_t := \mathrm{d}X_t \cdot \mathrm{d}X_t = \sigma - \limsup_{h\to 0^+}(X_{t+h} - X_t).$$

The above calculation rule is well defined, with respect to sequences, since if $Y_n \to Y$, with respect to the norm topology of L^p a subsequence of $(Y_n)_{n\in\mathbb{N}}$, which is denoted by $(Y_{k_n})_{n\in\mathbb{N}}$, such that $Y_{k_n} \to Y$, $\mathbb{P}$ a.e. Alike in the case of Itô calculus, higher class differentials' operations rely on the above essential rules in both cases of L^p-valued stochastic differential calculus and $\mathbb{R}$-valued stochastic differential calculus.

The analog of Itô Lemma in L^p is stated in the following way:

$$\mathrm{d}Y_t = \mathrm{d}f(t, X_t) = \frac{\partial f(t, X_t)}{\partial t}\mathrm{d}t + \frac{\partial f(t, X_t)}{\partial x}\mathrm{d}X_t + \frac{1}{2}\frac{f^2(t, X_t)}{\partial x^2}s_t\mathrm{d}t, \ (E),$$

where f is some $C^{1,2}$- function, which is $\mathbb{R}$-valued or the composition of such a function and some L^p-valued function, according to the corresponding stochastic differentials' definition. By $s_t dt$, we denote that s_t is actually the Radon–Nikodym derivative, with respect to the Lebesgue measure on $[0, T]$.

Remark 3.7 We avoid the definition of stochastic differential's calculation with respect to the **quadratic variation** process, since in the cases of L^1-valued stochastic processes, the quadratic variation process is not well defined. On the other hand, in both of the above ways of stochastic differentials' calculation rules, such a 'difficulty' does not exist.

4 Stochastic Differential Equations Under Riemann–Riesz Integration

A stochastic differential equation in both of the cases of stochastic differential calculus is stated in the form:

$$\mathrm{d}Y_t = a(t, X_t)\mathrm{d}t + b(t, X_t)\mathrm{d}X_t, \tag{1}$$

We directly may notice that such stochastic processes $(Y_t)_{t\in[0,T]}$ are the analog of Itô processes, with respect to some Brownian motion. The random variables Y_t for any $t \in [0, T]$ lie in $L^1(\Omega, \mathcal{F}_t, \mathbb{P})$, such that the corresponding integrals to be well defined.

Definition 4.1 A **strong solution** of the stochastic differential equation (1) is some stochastic process $(Y_t)_{t\in[0,T]}$, such that

$$Y_t = Y_0 + \int_0^t a(s, X_s)\mathrm{d}s + \int_0^t b(s, X_s)\mathrm{d}X_s, \mathbb{P}, a.e.$$

Definition 4.2 A **weak solution** of the stochastic differential equation (1) is some stochastic process $(Y_t)_{t\in[0,T]}$ such that

$$Y_t =^d Y_0 + \int_0^t a(s, X_s)\mathrm{d}s + \int_0^t b(s, X_s)\mathrm{d}X_s,$$

Alike in Brownian stochastic differential equations, a simple and useful example for the form (1), whose strong solution is provided in a certain form, is the linear stochastic differential equations. Their solution arise by applying the analog of Itô Lemma (E):

Definition 4.3 A stochastic differential equation of the form (1) is called **linear** if

$$a(t, X_t) = a_1(t)X_t + a_2(t),$$

$$b(t, X_t) = b_1(t)X_t + b_2(t),$$

where the domain a_1, a_2, b_1, b_2 is $[0, T]$ and their range is some non-empty subset of the real numbers.

We may notice that the (E)-Lemma implies a way for the specification of a strong solution, relying on the function $f : [0, T] \times \mathbb{R} \to \mathbb{R}$ such that $(t, x) \mapsto f(t, x)$.

5 Existence and Uniqueness of Strong Solution

Theorem 5.1 *If we consider the following conditions:*

$$|a(t, x) - a(t, y)| + |b(t, x) - b(t, y)| \leq K|x - y|, \ (1a),$$

where $|y| = y \vee (-y)$ denotes the absolute value, with respect to the vector lattice $L^1(\Omega, \mathcal{F}_T, \mathbb{P})$, where the values of $a(t, x), b(t, x), x$ are nonzero elements of $L^1(\Omega, \mathcal{F}_T, \mathbb{P})$, such that K is a positive real number. Then, a strong and unique solution to the stochastic differential equation of the form (1) does exist.

Proof We may define the operator $U : L^1(\Omega, \mathcal{F}_T, \mathbb{P}) \to L^1(\Omega, \mathcal{F}_T, \mathbb{P})$, in the following way:

$$U(X_t) = X_0 + \int_0^t a(s, X_s)ds + \int_0^t b(s, X_s)dX_s, \ \mathbb{P}, a.e.$$

By applying the **Banach contraction fixed-point theorem**, we obtain the conclusion, since $K > 0$. $X_0 = Y_0$, $\mathbb{P} - a.e.$, such that the expression of U to be compatible to the above definitions. We recall that since $L^1(\Omega, \mathcal{F}_T, \mathbb{P})$ is a **Banach Lattice**, then for any $x \in L^1(\Omega, \mathcal{F}_T, \mathbb{P})$, such that $|x| \leq |y|$, we obtain that $\|x\| \leq \|y\|$, where $\|.\|$ is the norm of $L^1(\Omega, \mathcal{F}_T, \mathbb{P})$. □

6 Appendix: Some Notions of Partially Ordered Vector Spaces

A vector space is called *partially ordered* by the cone E_+ if there exist some non-empty subsets E_+ of E, such that $E_+ + E_+ \subseteq E_+$, $\lambda \cdot E_+ \subseteq E_+$for any $\lambda \in \mathbb{R}_+$ (where $\cdot$ is the usual scalar product in E) and $E_+ \cap (-E_+) = \{0\}$. E_+ is called the *positive cone* of E, and the partially order relation induced by E_+ is defined as follows:

$$x \geq y \Leftrightarrow x - y \in E_+.$$

A Riesz space (vector lattice) E is *order complete* if every non-empty set of it, being order bounded from above, has a supremum and consequently any non-empty set of it, being order bounded from below, has an infimum. In the case of a vector lattice, $\sup\{x, y\}$ and $\inf\{x, y\}$ are denoted by $x \vee y, x \wedge y$, respectively. If so, $|x| = \sup\{x, -x\}$ is the *absolute value* of x. $x = x^+ - x^-$ for any $x \in L$, where L is a vector lattice. $x^+ = x \vee 0, x^- = (-x) \vee 0$. $L^1(\Omega, \mathcal{F}_T, \mathbb{P})$ is an order complete vector lattice. A detailed presentation for the above notions is contained in Aliprantis and Border (2006, Def.8.2), Aliprantis and Border (2006, Lem.8.4).

Conflict of interest Authors declare that there is not any conflict of interest concerning the present paper.

References

Aliprantis, C. D., & Border, K. C. (2006). *Infinite dimensional analysis, A Hitchhiker's guide* (3rd edn). Springer.

Floros, C., Gkillas, K., & Kountzakis, C. (2021). *An application of the Arrhenius equation in portfolio modeling available at SSRN*. https://doi.org/10.2139/ssrn.3806118

Wong, E., & Zakai, M. (1965). On the relation between ordinary and stochastic differential equations. *International Journal of Engineering Science, 3*, 213–229.

A Machine Learning Approach to Entrepreneurial Finance Modelling

Max Berre

Abstract Traditionally, estimating valuation relies on firm data and concrete economic indicators. So does modelling of startup investment selection and startup survivability. However, recent advancements in machine learning have given rise to customizable segmented-modelling approaches. While classical economic theory describes that firm valuations and survival rates are modelled based on revenues, growth rates, and risk, the valuation of startup often proves the exception to the rule. Meanwhile both startup investor selection and startup valuations are influenced by revenues, risks, age, and macroeconomic conditions, specific causality is traditionally a black box. Likewise, for startup survivability, which is known to be influenced by risks, revenues, age-of-firm, and access to finance, specific causality is also unclear. Because details are not disclosed, roles played by other factors (industry, business models, geography, and intellectual property) can often only be guessed at. This study is an in-depth examination outlining methods and approaches for application of segmented modelling in entrepreneurial finance, as well as ways in which they can be applied using existing data for purposes to examine selection, valuation, and survivability.

Keywords CART · Decision tree · Valuation · Startup valuation · Startup selection · Investment selection · Startup survival startup survivability · Venture capital · Entrepreneurial finance · Machine learning · Hierarchical analysis

M. Berre (✉)
Audencia Business School, Nantes, France
e-mail: mberre@audencia.com

Nyenrode Business University, Amsterdam, Netherlands

C. Zopounidis et al. (eds.), *Operational Research Methods in Business, Finance and Economics*, Lecture Notes in Operations Research,
https://doi.org/10.1007/978-3-031-31241-0_2

1 Introduction

Why are startup in Massachusetts-based startup substantially more likely to survive into their fourth year than those in New Hampshire? Why do startup in London attract higher valuations than those in Paris, Berlin, or Milan, even when they are based in similarly sized economies, share the same industries and many of the same investors? Why do healthcare-industry startup work their way through the VC-selection deal funnel more efficiently than IT-industry startup?

Whereas classical economic theory describes that firm valuations and survival rates are modelled based on revenues, growth rates, and risk, valuation of startup often proves the exception to the rule with startup differing significantly in terms of information environment, time structure of transactions, and linkages between investors and investees (Bellavitis et al. 2017). Given their specific characteristics, startup are notorious for being difficult to value, while their investment selection is difficult to predict, and their survivability can vary dramatically across industry (Damodaran, 2009) and across geography (Gonzalez, 2017). These difficulties are driven by opacity of both startup and investors, as well as short histories, and complex intangible assets held by startup (Damodaran, 2009), as well as disruption potential (Damodaran, 2019).

Overall, this is intended how to guide outlining approaches and methods for application of segmented hierarchical modelling in entrepreneurial finance, as well as how they can be applied using existing data and regression models.

The rest of this study proceeds as follows: the subsequent section describes segmented models, while also describing where they have appeared in both practitioner-focused grey literature, as well as in peer-review literature. After this, Sect. 3 describes how segmented models can be made hierarchical, as well as describing how they can be used for microtargeting-based approaches. Lastly, the discussion and conclusion section outlines why segmented, hierarchical, and microtargeting approaches are used by industry practitioners, by describing their added-value vis-à-vis more traditional approaches.

2 Why Segmented Models: What Do We Aggregate?

A relatively widespread theoretical approach used typically for both startup selection and startup valuation is that of the scorecard-based approach. Given that Fama (1970) describes factors as information-subsets which have the potential to drive price-signals, and which can range from historical-values to disclosures and privileged-information, their incorporation may be highly relevant. The primary advantage of the scorecard approach is the ability to incorporate qualitative, geographic, sectoral, or categorical determinants in several ways. In entrepreneurial finance, these factors might range from non-financial and deal characteristics prevalent in given sectoral or municipal ecosystems, the role of national-level or market-condition determinants, to the role that business models or ownership structures and legal form may play in survivorship or investment selection. This approach can be used to estimate either valuation or selection ranking and is capable of establishing insights even as detailed related economic and financial information is missing, scare, or unevenly available.

Segmented models are modular and relatively straightforward model approaches based on summation of market conditions, key characteristics, and deal conditions developed primarily by industry practitioners. One critical advantage of this sort of model approach is that valuation, selection, or survival probability can be modelled, captured, and understood while including specific categorical information, which can be both general, highly specific, and/or be organized as joint, combined, or hierarchical segmentation.

Mechanically speaking, reliance on optimal arrangement of multiple decision factors is central to model functionality. In parallel, multiple-criteria decision analysis (MCDA), which explicitly evaluates conflicting-criteria in decision-making is described Zopounidis et al. (2015) as being used for portfolio and investment evaluation and selection, is usually implemented in terms of fundamental factors. While valuation factors do not necessarily conflict, valuation impacts of trade-offs and fault lines may constitute important model elements which need to be taken into consideration.

In a similar vein, Zopounidis and Doumpos (2002), who examine and model investment selection, describe that choices of investment project are strategic decisions made by human agents (e.g. financial managers or venture capitalists) and not by the model; the decision makers become more and more deeply involved in the decision-making process. Citing this, Dhochak and Doliya (2019) outline apply fuzzy analytic hierarchy process to startup valuation, claiming that fuzzy AHP is a well-suited methodology to evaluate startup valuation due to close resemblance to cognitive human decision-making approaches.

Ellis et al. (2001) meanwhile make use of a segmented multi-criteria modelling approach in order to model supplier success in meeting customer expectations in the high-technology marine equipment industry. To demonstrate divergent views between shipbuilders and shipowners in both the European and Japanese markets. Figure 1 outlines ranked differences in supplier success factors, finding that Japan's shipbuilder market places higher importance on prices than do European shipbuilder markets, while Japan's shipowner market places lower importance on maintenance than do European shipowner markets.

2.1 *Practitioners: Segmented Models in Markets*

In industry, scorecard approaches are typically used by business angels. Scorecard valuation approaches which have emerged from industry prominently include Payne (2011) and Berkus (2016). Perhaps the most well-known segmented startup valuation model is the scorecard model, outlined by Payne (2011). Outlined in Table 1,

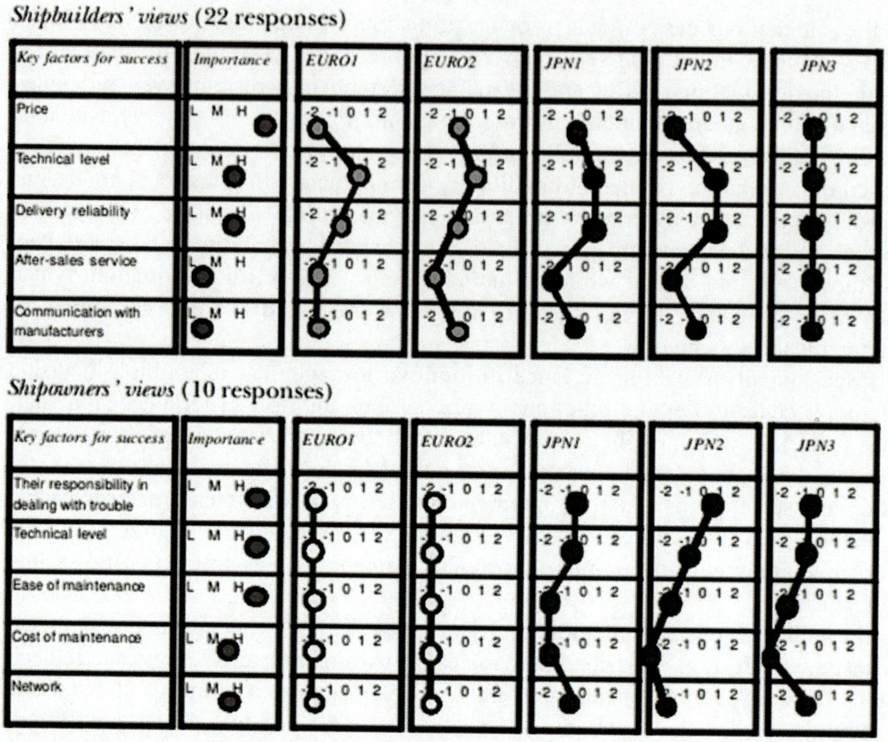

Fig. 1 Differentiated segmental positioning modelling supplier success in marine high-tech equipment. *Source* Ellis et al. (2001)

Payne's scorecard model segments valuation into management team, target market, competitive environment, and further funding need.

Focusing specifically on valuation, a well-known alternative to the scorecard model is the Berkus model (Ernst & Young, 2020). Outlined in Fig. 2, the Berkus model segments valuation into component risks.

2.2 Segmented Models in Peer Review Literature

Meanwhile, in published economic literature, this same concept emerges as summation-based valuation models. Prominent examples include models published by Hand (2005), Miloud and Cabrol (2011), and Sievers et al. (2013). For example, Eq. 1 describes Hand (2005)'s startup-valuation model, which is driven by deterministic valuation factors segmented into financial-statement data such as assets, Net Income, and cash flows, on one hand, and operational and industry-related data on the other.

Table 1 Abbreviated Payne scorecard model

Weighting		Impact on startup selection and valuation
0–30%	Impact	*Strength of the entrepreneur and the management team*
	+	Many years of business experience
	++	Experience in this business sector
	+++	Experience as a CEO
	++	Experience as a CFO, COO, or CTO
	+	Experience as a product manager
	–	Experience in sales or technology
	– –	No business experience
0–25%	Impact	*Size of the opportunity*
		Size of the target market (total sales)
	– –	< $50 million
	+	$100 million
	++	> $100 million impact
		Potential for revenues of target company in five years
	– –	< $20 million
	++	$20–$50 million to > $100 million (will require significant additional funding)
0–15%	Impact	*Strength of products and intellectual property*
	– – –	Not well defined, still looking for prototypes
	0	Well defined, prototype looks interesting
	++	Good feedback from potential customers
	+++	Customer orders or early sales
0–10%	Impact	*Competitive environment*
		Strength of competitors in this marketplace
	– –	Dominated by a single large player
	–	Dominated by several players
	++	Fractured, many small players
		Strength of competitive products
	– –	Competitive products are substantial
	++	Competitive products are weak
0–10%	Impact	*Marketing/sales/partners*
		Impact sales channels, sales, and marketing partners
	– – –	Have not discussed sales channels
	++	Key beta testers identified and contacted
	+++	Channels secure, customers placed trial orders
	– –	No partners identified

(continued)

Table 1 (continued)

Weighting		Impact on startup selection and valuation
	++	Key partners in place
0–5%	Impact	*Need for additional rounds of funding*
	+++	None
	0	Additional angel round
	−−	Need venture capital

Source Ernst and Young (2020)

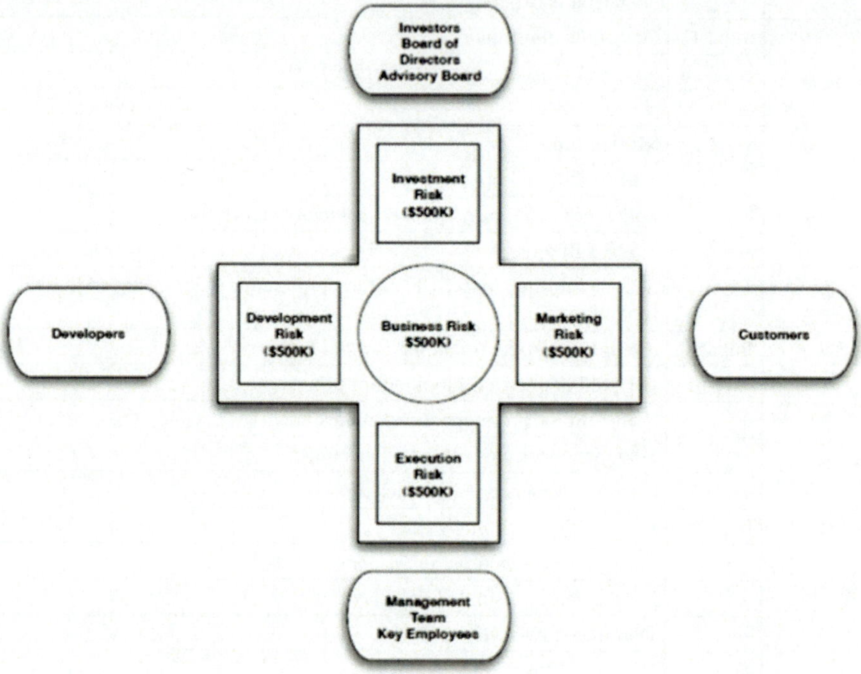

Fig. 2 Berkus model for startup valuation. *Source* Berkus (2016)

Equation 1 Hand (2005) Summation-based segmented valuation model

$$\text{Hand (2005) Ln(Pre-Money Valuation)} = \sum \theta_{\text{b}}\text{Ln(Financial Statement Data}_{bik}\text{)} + \sum \Upsilon_{\text{c}}\text{Ln(NonFinancial Statement information}_{cik}\text{)} + \varepsilon_{ik} \quad (1)$$

Meanwhile, Eq. 2, another prominent segmented startup-valuation model outlines the Sievers et al. (2013) summation-based valuation model, describing valuation on the basis of summation of financial, and non-financial firm attributes, as well as deal characteristics and relevant valuation coefficients. Essentially, whereas Hand (2005) segments valuation factors into accounting and non-accounting data, Sievers et al. (2013) segment valuation factors into financial factors such as revenues, risks, or capital invested, and non-financial factors including operational and industry-level data, and deal characteristics such as investor syndication, and investment-deal clauses such as redemption, tag-along, and ratchet clauses in the investment deal.

Equation 2 Sievers et al. (2013) Summation-based segmented valuation model

$$\log(\text{Valuation}_{it}) = \sum \Phi\text{Non-financial}_{it} + \sum \Delta\text{Financial}_{it} + \sum \Psi\text{Deal Characteristics}_{it} \tag{2}$$

Sievers' model estimates valuation by summation of the established segments, as is the case with the Berkus and Payne models. Overlooked, however, are interactions and hierarchies among valuation determinants.

Mechanically speaking, an alternate functional form to express segmented valuation can be elaborated via the staged valuation approach. Examples of this approach include the Startup Valuation Meta-Model developed by Berre and Le Pendeven (2022) outlined in Eq. 3. In addition to valuation factors themselves, this approach accounts for phases, interactions, and hierarchies among valuation factors. Formally:

Equation 3 Berre–Le Pendeven (2022) Valuation meta-model for startup

$$\text{Pre-Money Valuation} = f\left(\left(\left(\sum \text{Start-Up Value}\right)\sum \text{Deal Value}\right)\sum \text{Deal Valuation}\right) \tag{3}$$

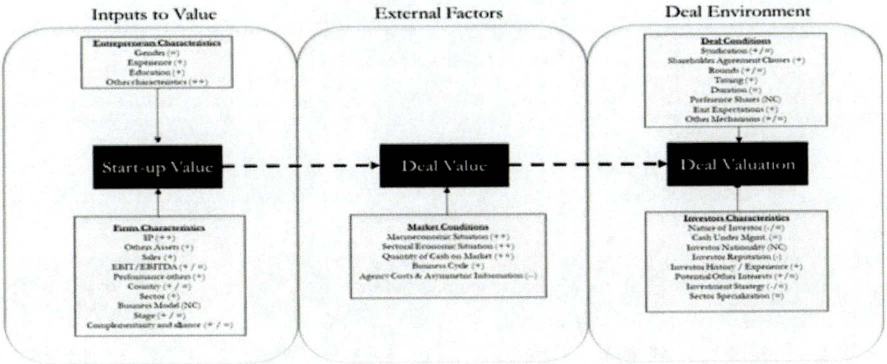

3 From Segmentation to Hierarchical Microtargeting Models

Recently, the emergence and development of supervised machine learning techniques has led to increasing methodological sophistication of scorecard approaches, as predictive techniques incorporating categorical, qualitative, geo-spatial, and ordinal data have become increasingly widespread.

Mechanically speaking, microtargeting by means of datamining is described in detail by Murray and Scime (2010), as the process of inductively analysing data to find actionable patterns, fault lines, and relationships within the data, on the basis of trends drawn from both numerical and descriptive characteristics, such as average family age, family composition, and geographic area, via construction of decision trees, an analytical technique which is both explanatory and predictive, and which is used for both variable predictions, as well as to provide specific insights concerning structure, segmentation, and interrelationships among data.

This approach grants insight into how specifically any outcome variable's value is dependent on the model's deterministic factors, with each identifiable fault line constituting segments of individuals. Fundamentally, microtargeting by means of data mining can allow scorecard-based modelling approaches to incorporate qualitative and categorical data hierarchically, to a potentially extreme degree of detail.

Functionally speaking, a hierarchically structured model resulting from a microtargeting approach can be expressed via a staged model approach. For instance, Fig. 3 displays the architectural form that the Berre–Le Pendeven Startup Valuation Meta-Model described in Eq. 2 would adopt, expressed as a hierarchical decision tree.

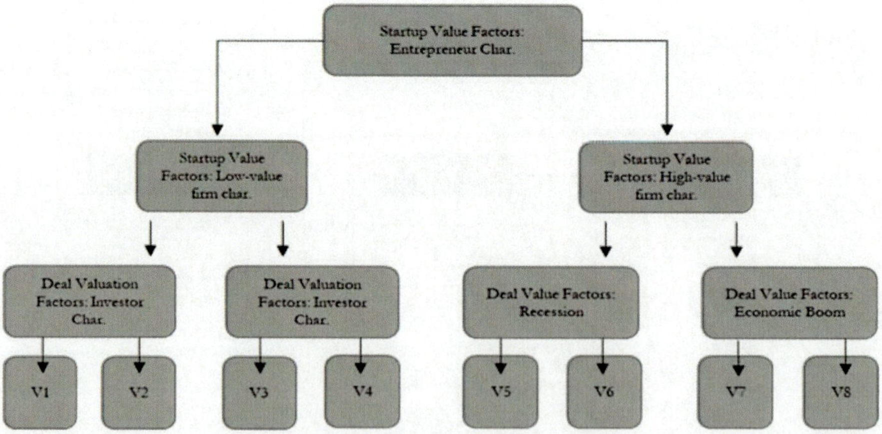

Fig. 3 Decision tree based on the Berre–Le Pendeven meta-model

3.1 Hierarchical Approaches: Regression Trees and Random Forests

Functionally speaking, CART-based microtargeting using regression tree and Random Forest approaches, the latter of which agglomerate large numbers of regression trees, can reorganize determinant impacts causing several key insights to emerge, which might otherwise be missed by regression-model approaches, or by more rudimentary estimation models. Firstly, key fault lines are expressed as threshold values along which branches diverge. Secondly, qualitative or categorical determinant factors such as geographical, sectoral, and business-model data, which has the potential to be information dense are taken into account. Thirdly, CART trees demonstrate areas and subsections of the data where given valuation determinants might be more or less influential, granting very precise insight into how valuation emerges.

For empirical-model estimation purposes, the informational content of descriptive and categorical characteristics such as geography, industry, legal form, or business model are often overlooked, despite the general possibility that these characteristics might bring-to-bear explanatory power equivalent to multiple associated numerical variables. Meanwhile, use of fixed effects to incorporate descriptive categorical characteristics suffers losses in explanatory power as the number of descriptive characteristics increases Wooldridge (2010), whereas microtargeting approaches improve their accuracy as the number and density of these characteristics increases.

Consequently, a key advantage of this approach is that startup valuation, selection, or survival probability can be microtargeted by including ever smaller and more specific categorical information, or combinations thereof.

3.2 Functional Form of Segmented Models

According to Krzywinski and Altman (2017), a CART approach does not develop or express a prediction equation. Instead, this approach partitions data along predictor axes into subsets with homogeneous values of the dependent variable. This in mind, machine learning algorithm reliance on optimal arrangement of decision factors is central to model functionality.

In tree-based approaches, this process is represented by decision trees, which can be used to make predictions from new observations. Several functional-form options exist mathematically, which are used operationally by practitioners in markets or in research settings. Furthermore, the combination and/or selective use of these can be a useful way to investigate and model causal relationships in detail. Within entrepreneurship studies, this can be used to investor selection, startup valuation, and startup survivability.

Log Transformation

Because log transformation renders summation and multiplication interchangeable, the use of variable log transformation can dramatically simplify regression models and mathematical relationships for the purposes of empirical specification (Benoit, 2011), while also lending themselves to model flexibility. Given the product property of logarithms, it is possible to express the model in its entirety as a summation model for intermediate-stage purposes, given the interchangeability of logarithm multiplication and summation (Miller et al., 2010). Specifically, this means that intermediate-stage model functional forms can reoriented in terms of variable order and in terms of interaction effects.

Moreover, log transformation "flattens" empirical relationships, by restraining the effect outliers have on variable medians and means. Because regression trees and partitioning methods in general are sensitive to outlier influence from dependent-variable outliers (Khan et al., 2013), flattening of outliers has potential to substantially increase explanatory power to regression-tree models, as log transformation reduces estimation problems associated with percentage changes from baseline (Keene, 1995), while maximizing data-scale-flattening (Ribeiro-Oliveira et al., 2018). Variables showing skewed distribution can also be made symmetric using log transformation (Keene, 1995).

Conversely, since log transformation also impacts multiplicative models (Benoit, 2011), the particular architectural shape of functions being modelled becomes unclear, as multiplication, summation, ratios, and other functional-form elements might also become unclear.

To reach a viable final outlook, one would need to see the model's log-transformed expression alongside the original expression, whose functional form captures in detail both variable order and possible interaction terms. In order to establish a tree, however, both variable order and relative variable importance need to be established. Overall, interaction terms between and among regression variables can shed some light on how specifically the model's explanatory variables interact with each other. This may indicate within-tree variable position, granting a more holistic and complete view on relationship and model causality structures.

Regression-Model Equations

Functionally speaking, regression-model equations, which consist of a summation of key variables, modified by factor coefficients, alongside constants and error terms. Fundamentally, this layout structure lends itself to near-direct transposition of segmented valuation approaches, was well as the approximation of most classically established firm valuation models, ranging from discountedcash flow valuation (DCF) approaches, to multiples-valuation approaches.

Because regression-model equations are generally expressed as summation functions, with each of the model's terms consisting of a variable and a coefficient, valuations can essentially be expressed as a summation of variables, coefficients, constants, and error terms. For instance, a discounted-revenue-based valuation, incorporating similar information to a discounted cash flow valuation (DCF), could approximate a free cash flow to equity (FCFE) approach by regressing valuation on present and historic Net Income figures in order to capture both free cash flow and its growth rate, as well as risk factors which drive the discount rate, which can be expressed as combinations of the risk-free rate and the applicable equity risk premia described in the CAPM model. This is modelled in Eq. 4.

Equation 4 Valuation regression model simulating free cash flow to equity (FCFE)

$$\begin{aligned}\text{Valuation}_{it} &= \alpha_i + \beta_1(\text{Net Income}_{it}) + \beta_2(\text{Net Income}_{it-n}) \\ &\quad + \beta_3(\text{Risk-Free rate}_t) + \beta_4(\text{Risk-Premium}_{it}) + u_{it}\end{aligned} \tag{4}$$

Meanwhile, a multiples-valuation approach, whose widespread popularity flows from its simplicity and ease of communication, as well as its ability to communicate the market's current mood (Damodaran, 2002), might seek to estimate valuation from as few as one valuation factor drawn from either a firm's balance sheet, income statement, or statement of cash flows. This however may come at the cost of sample selection, as developing a sample of relative firms and assets against which to compare valuation, can lead to standardization (or assumption of standardization) of variables outside of the valuation model. According to Damodaran (2002), the most widespread multiples-valuation model is the price/sales ratio, which describes valuation as a function of sales revenue, as outlined in Eq. 5

Equation 5 Price-to-sales ratio

$$\text{Price-to-Sales Ratio} = \frac{(\text{Firm's Total Market Share} - \text{Price})}{\text{Sales Revenue}} \tag{5}$$

Equation 6 demonstrates this ratio as an ordinary-least-squares (OLS) regression model, given by the parameter sales revenue, while β estimates price-to-sales ratio, whereas outside factors ranging from quantitative valuation determinants such as borrowing costs, R&D, CAPEX, or total assets (or asset subsets such as IP assets), to qualitative valuation factors such as those driven by industry or economic geography are sample selected to be constant, or assumed to be constant.

Equation 6 Price-to-sales ratio as an OLS regression model

$$\text{Valuation}_i = \alpha_c + \beta_c(\text{Sales Revenue}_i) + u_i \tag{6}$$

Apart from the use of regression-model functional form to express classical models, the OLS regression-model's functional form can also be used for summation-based segmented models, such as those outlined in Eqs. 1 and 2. In fact, this is even the case for models using hierarchical approaches, such as Mahmoud et al. (2022) express random-forest regressions using regression-model equations, simulating the summation-based segmented functional form of an OLS model.

Decision-Tree Model Functional Forms

Overall, substantial flexibility exists concerning the various functional forms that decision-tree models could conceivably adopt considering contexts in which they could be deployed, factors enumerated by the model, and both their relative and hierarchical explanatory power. While Krzywinski and Altman (2017) describe that a CART approach does not develop a prediction equation, CART regression-tree results can be used to modify or extend segmented models. Fundamentally, regression-tree model outputs make possible two practically viable segmentation model approaches.

Mahmoud et al. (2022), for example, express random-forest regressions using regression-model equations, simulating the functional form of an ordinary-least-squares model. This modelling approach has the advantage of capturing the causal relationship's overall directionality, which can be tested empirically, without specifically precluding existence of complex model functional forms.

Comparing Model Goodness-of-Fit

In principle, accuracy of regression-tree models can be compared to those of equivalently constructed regression models on the basis of their respective goodness-of-fit indicators. Whereas linear regressions are typically evaluated on the basis of R^2, Sandeep (2014) and Firmin (2021) outline that regression trees should be evaluated on the basis of $1 - R^2$ root-mean-squared-error.

Weighted Summation Segmentation

First, a rudimentary "back-of-the-envelope" approach to segmentation can be considered to be a modification of the Payne scorecard model, which includes model weighting to its segmentation approach. In order to obtain regression-tree model weights from the CART approach, it suffices to examine the model's variable-importance scores. While CART variable-importance outputs can aggregate to a maximum of 100%, as is the case in Table 3, aggregate variable-importance model outputs might also aggregate to less than 100%. While for CART models whose aggregate variable importance adds to 100%, it would suffice to assign variable-importance figures as weighting coefficients. For instances in which observed variable-importance outputs aggregate to less than 100%, however, factor-importance proportionality would need to be calculated as a first step, as outlined in Eq. 7:

Equation 7 CART variable-importance proportionality

$$\text{Factor-Coefficient}_i = \sigma_i(X)_i = \frac{\text{Variable Importance}_i}{\sum_n^i \text{Variable Importance}_i} \tag{7}$$

Fundamentally, this approach is useful as a generally applicable model approach, yielding a Payne-like scorecard model, which can be applied in a general fashion to entrepreneurial and startup markets at-large. For example, a Payne-style scorecard model, involving model weights, which could be constructed on the basis of firm characteristics and market characteristics, can take the form outlined in Eq. 8, combining the FCFE valuation factors with Payne model factors outlined in Table 1:

Equation 8 Weighted summation segmentation regression-tree valuation model simulating the FCFE valuation approach

$$\begin{aligned}\text{Valuation}_i &= \sigma_1\beta_1(\text{Net Income}_i) + \sigma_2\beta_2(\text{Risk-Free rate}_i)\\ &\quad + \sigma_3\beta_3(\text{Risk-Prem.}_i) + \sigma_4\beta_4\big(\text{Size of Opportunity}_i\big)\\ &\quad + \sigma_5\beta_5(\text{Competitive Env.}_i) + \sigma_6\beta_6(\text{IP}_i)\end{aligned} \tag{8}$$

where

$$\sum_{i=1}^{n} \sigma_i = 1 \quad \text{but we observe} \quad \sum_{i=1}^{n} \widehat{\sigma_i} \leq 1$$

Here, σ refers to the weighting coefficient n of startup i, driven by variable importance (for example, the scale of Net Income's impact on startup i's valuation), while β refers to the impact coefficient n of startup i (for example, risk premium is a valuation determinant known be a constituent factor of the DCF-model discount rate (Damodaran, 2009), and as such, can be expected to have a negative β-coefficient).

Mechanically, this approach is viable for either continuous numerical variables, such as those drawn from a firm's financial statements (i.e. Net Income, fixed assets, etc.), as well as market data (i.e. business cycle and macroeconomic indicators), or for categorical and binary variables such as entrepreneur characteristics or intellectual property. Additionally, because CART regressions partition data along the predictor axes into dichotomous subsets, categorical variables (i.e. classifications such as sectoral-industry classifications and business-model classifications, as well as economic-geography variables such as counties, cities, inclusion in regional clusters) which are treated as binary variables.

Hierarchical Ordinal Segmentation

A second modelling approach can be referred to as the hierarchical ordinal segmentation approach. Given that the data are partitioned along predictor axes into subsets with homogeneous values of the dependent variable, a more complex hierarchical approach is also plausible. The basis of this approach would begin with adoption of terminal-node average values as ω-coefficients. These can subsequently be multiplied by a regression-tree's branch conditions and branch thresholds, as follows:

$$\omega_i(X)_j\left(\begin{cases} = 1 \text{ if } X \text{ is true} \\ = 0 \text{ if } X \text{ is false}\end{cases}\right.$$

Or

$$\omega(X)_j\left(\begin{cases} = 1 \text{ if } X \text{ is above the threshold} \\ = 0 \text{ if } X \text{ is below the threshold}\end{cases}\right.$$

Subsequently, the regression-tree model can be elaborated for any specific startup in accordance with the position it occupies in the regression tree. Equation 9 describes this functional form.

Equation 9 Valuation regression-tree model using hierarchical ordinal segmentation

$$\text{Valuation}_i = \omega_i \left(\prod_{i1}^{in} \text{Branch Threshold}_i \right)_i + \cdots + \omega_n \left(\prod_{n1}^{nn} \text{Branch Threshold}_n \right)_n \tag{9}$$

As a specific example building on Eq. 9, establishing a specific valuation model, Eq. 10 applies the hierarchical ordinal segmentation approach to the combined FCFE-market conditions valuation model outlined in Eq. 8 and ranking the nodes in hierarchical order following their order in Eq. 8. Note that this causes their order stated in the equation to change somewhat to reflect the conditionality relationship.

Equation 10 Valuation regression tree using hierarchical ordinal segmentation model approach

$$\begin{aligned} \text{Valuation}_i = {} & \omega_i \left(\prod_{i}^{I} \text{Net Income}_i \right)_i + \omega_j \left(\prod_{j}^{J} \text{Risk-Free rate}_j \right)_j \\ & + \omega_k \left(\prod_{k}^{K} \text{Risk-Premium}_k \right)_k + \omega_l \left(\prod_{l}^{L} \text{Size of Opportunity}_l \right)_l \\ & + \omega_m \left(\prod_{m}^{M} \text{Competitive Env.}_m \right)_m + \omega_n \left(\prod_{n}^{N} \text{IP}_n \right)_n \end{aligned} \tag{10}$$

Fundamentally, a key difference between this approach and the weighted-summation approach is that this approach is specific to the individual startup's position within the decision tree. Essentially, this means that the segmentation's functional form differs from that of weighted-summation approach, since a startup's placement on the regression tree may indicate functional form featuring either the repetition or omission of some of the regression-model's valuation determinants.

Another essential difference between the approaches is that while the weighted-summation approach can grant a holistic view of σ-weights across the dataset as a whole, the ordinal-model approach can directly provide a valuation estimate by placing the firm along regression-tree's terminal nodes (i.e. the regression-tree's leaf nodes).

Two-Tiered Approach

Given that inclusion of categorical variables has the potential to unearth valuable informational insights of both qualitative and quantitative nature and has the potential to be as information dense as the joint inclusion of multiple numerical variables, their use for research purposes remains a very valuable tool (Neter et al. 1990; Wooldridge, 2010). This is in particular the case with fixed-effects regressions, given that they can meaningfully incorporate categorical indicators such as geographical or industry-level designations. In the face of multiple information-dense categorical variables, however, this approach is subject to a hard limit in that explanatory power of joint fixed effects can be limited as the number of categorical variables grows.

What this means therefore is that either OLS or fixed-effects regressions can be deployed in order to capture the general causal overview among the model determinants and in order to detect information density and explanatory power of applicable categorical labels. In order to elaborate on any OLS or fixed-effects findings, CART (or possibly other cluster-driven approaches) can be utilized with the aim of enrichment or corroboration of findings.

Taking this into consideration, combined approaches are possible, with the potential to outperform single-method analysis in two important ways. Firstly, this approach can outperform a stand-alone OLS-based summation approach because the two-tiered approach can grant insights on the role, hierarchy, and relative-position of the model's near-significant explanatory factors (i.e. near-significant factors often have regions or subsets of the data, for which they are significant). Secondly, two-tiered approaches can provide detailed insight vis-à-vis scale and sign of factor impacts (i.e. β-coefficients), thereby improving upon stand-alone CART-based weighted summations.

4 Modelling Investment Selection and Startup Survivability

A Segmented Approach to Selection

Aside from predicting and modelling valuation, machine learning-driven segmented models also have viable applicability for modelling startup selection and startup survivability, both of which are parallel entrepreneurial finance topics which have historically encountered modelling difficulties. Similar to startup valuation, irregularity and non-transparency of data constitute considerable obstacles to model accuracy (Damodaran, 2009).

Startup selection, while a very nearby parallel entrepreneurial finance topic, which shares many of the same prominent authors, faces the additional difficulty of qualitative and intangible factor determinants and decision criteria assuming a more widespread and prominent role among business angel and venture capital investors responsible for startup-selection decisions. A segmented-model approach to startup-selection decisions faced by venture capitalists and business angels is presented by Siskos and Zopounidis (1987), which includes both decision weights and ordinal rank, as outlined in Table 2 (Siskos & Zopounidis, 1987).

Table 2 Weighting of marginal utilities for VC investment decision

Rank	Criteria	1st analysis weight	2nd analysis weight
1	Information security	0.044	0.095
2	Market trends	0.000	0.005
3	Market niche/position	0.164	0.162
4	Conjuncture sensibility	0.009	0.085
5	Result trends	0.347	0.167
6	Expected dividend rate	0.031	0.107
7	Quality of management	0.031	0.247
8	Research and development level	0.000	0.000
9	Accessibility to financial markets	0.373	0.132

Siskos and Zopounidis (1987). http://linkinghub.elsevier.com/retrieve/pii/0377221787900403

Fundamentally, while Siskos and Zopounidis (1987) use ordinal regression analysis to reach Table 2's findings, these results provide sufficient detail for the construction of a decision-tree model equation, using a weighted-summation functional form.

Citing Siskos and Zopounidis (1987)'s ordinal regression analysis approach, Dhochak and Doliya (2019) expresses ordinal selection data via fuzzy analytic hierarchy process (Fuzzy AHP), outlined in Fig. 4, consisting of decision criteria and decision sub-criteria. Essentially, the model's hierarchy represents cognitive organization, dividing criteria into various types of firm-level internal-resources, industry-level resources, and network effects.

Building on Table 2's multi-criteria decision factors, Fig. 5 proposes a hierarchical decision tree for selection ranking based on Siskos and Zopounidis' top five decision criteria, prioritizing the dominant decision criteria, according to their analysis weights, with accessibility to financial market and result trend constituting the top decision-tree branches, while market niche/position appears in multiple lower branches, due mainly to its heavy analysis-weight score in both first and second analysis. Ultimately however, HCA, CART or Random Forest results would provide the specific functional form for the final decision tree. In contrast to the Dhochak and Doliya fuzzy AHP approach, Fig. 5's decision-tree approach arranges hierarchy according to likely explanatory power, rather than cognitive factor * organization levels.

This approach may be especially useful, in particular to analyse choice models, which incorporate or suggest qualitative data points, such as the entrepreneur personality traits outlined in Murnieks et al. (2016), or the entrepreneur and investor traits described in Andreoli (2022).

A Segmented Approach to Survivability

Startup survival, on the other hand, would adopting a functional form requiring a binary dependent variable and can draw on parallels from conditional probability

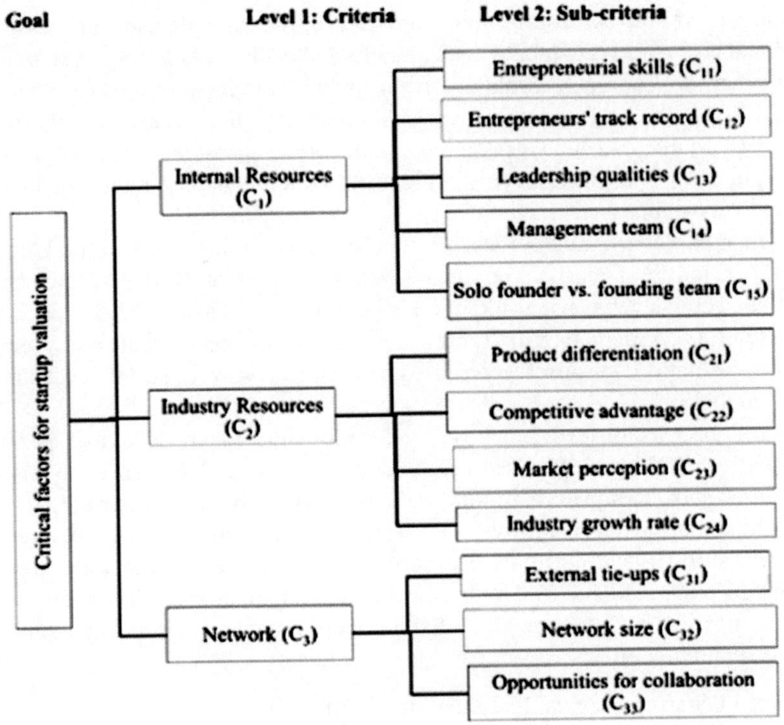

Fig. 4 Dhochak and Doliya startup-selection hierarchical decision model using fuzzy AHP

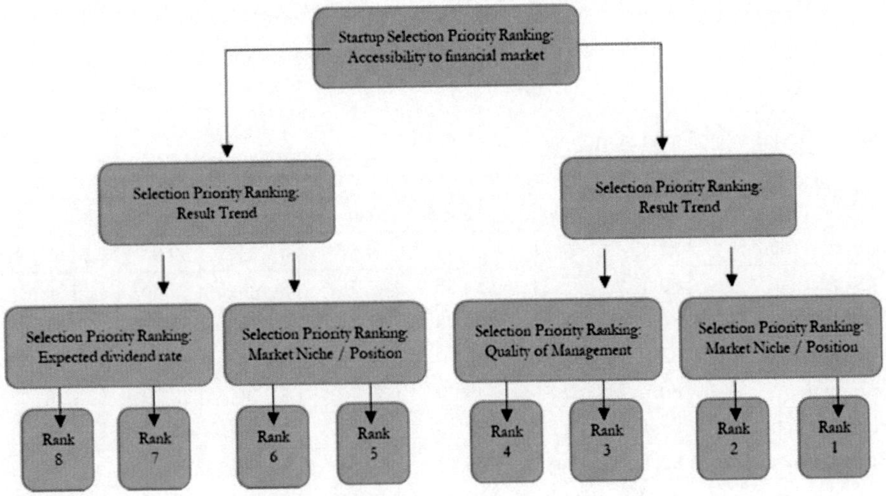

Fig. 5 Startup-selection ranking decision tree based on Siskos and Zopounidis decision weights

modelling, used in stress testing. Rebonato (2010), for example uses progression of conditional probabilities to model bank defaults and bank stress testing. A segmented machine learning approach would involve using a hierarchical tree-based algorithm (e.g. CART, ACH, or Random Forest). Alternatively, the tree-based algorithm's dependent variable can be expressed as a categorical variable instead of a binary variable, in order to capture varying degrees of financial distress, rather than simply bankruptcy as a binary term.

Specifically, determinants of startup survivability are described in the literature primarily in terms of market conditions. While Damodaran (2009) draws on the post-1998 sector-level survival likelihoods compiled by Knaup (2005) and Knaup and Piazza (2007), displayed in Table 3, in order to estimate credit-risk premium for startup valuation, purposes, it can also be used to construct sector-level startup-survival modelling.

In addition to sector-level survivability-determinants, macro-level market conditions such as GDP growth rates, prime-lending rates, and presence of business accelerators and startup accelerators are also known to play deterministic roles in modelling startup survivability (Gonzalez, 2017). In addition, Gonzalez (2017), which draws on US state-level data, describes considerable sectoral and state-level variation in one-year and four-year startup-survival likelihood. Econometrically, these findings can be represented as fixed-effects model including both state and industry-level fixed effects, as per Eq. 11.

Equation 11 Startup-survival likelihood fixed-effects model

$$\begin{aligned}\text{Survival Likelihood}_{\text{State, Industry}} = {} & \beta_1 \left(\text{Real GDP Growth}_{\text{State, Industry}}\right) \\ & + \beta_2 \left(\text{Prime Interest Rates}_{\text{State, Industry}}\right) \\ & + \beta_3 \left(\text{Accelerators}_{\text{State, Industry}}\right) + \varepsilon_{\text{State, Industry}} \end{aligned} \quad (11)$$

Table 3 Sector-level 7-year startup-survival likelihood

	Proportion of firms that were started in 1998 that survived through (%)						
	Year 1	Year 2	Year 3	Year 4	Year 5	Year 6	Year 7
Natural resources	82.33	69.54	59.41	49.56	43.43	39.96	36.68
Construction	80.69	65.73	53.56	42.59	36.96	33.36	29.96
Manufacturing	84.19	68.67	56.98	47.41	40.88	37.03	33.91
Transportation	82.58	66.82	54.70	44.68	38.21	34.12	31.02
Information	80.75	62.85	49.49	37.70	31.24	28.29	24.78
Financial activities	84.09	69.57	58.56	49.24	43.93	40.34	36.90
Business services	82.32	66.82	55.13	44.28	38.11	34.46	31.08
Health services	85.59	72.83	63.73	55.37	50.09	46.47	43.71
Leisure	81.15	64.99	53.61	43.76	38.11	34.54	31.40
Other services	80.72	64.81	53.32	43.88	37.05	32.33	28.77
All firm	81.24	65.77	54.29	44.36	38.29	34.44	31.18

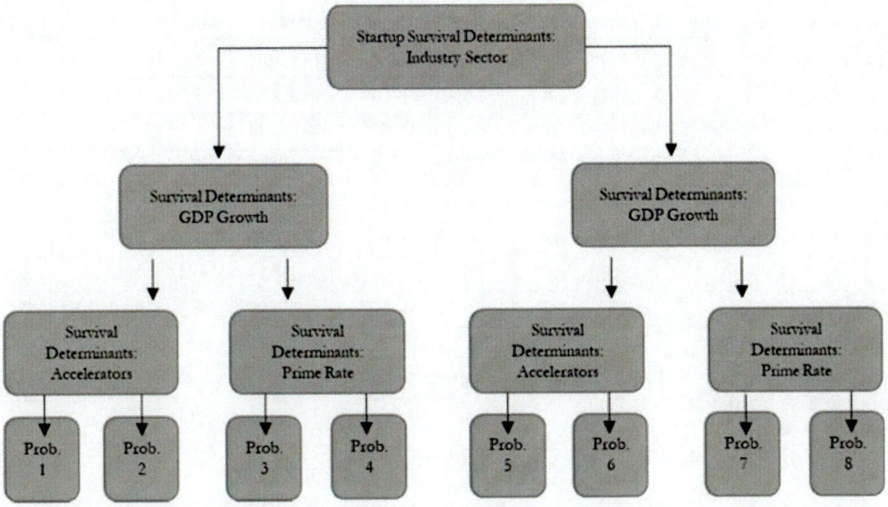

Fig. 6 Survival likelihood decision tree based on Damodaran (2009) and Gonzalez (2017)

Alternately, modelling startup survivability as a hierarchical decision-tree model can incorporate both the numerical determinants driving the startup-survival likelihood model outlined in Eq. 11, as well as the categorical variables which are used to construct Eq. 11's fixed effects. Hierarchically, Fig. 6 captures this relationship, adding the proposition that high-GDP growth startup survivability may by more influenced by prime-lending rates, whereas low-GDP growth startup survivability may by more influenced by presences and accessibility of startup accelerators. Because industry-level effects are described by both Damodaran (2009) and Gonzalez (2017), they are prioritized in this model, as they likely have substantial explanatory power.

5 Example of CART-Based Microtargeting Valuation Using a Single Categorical Variable

Tables 4 and 5 demonstrate the OLS and CART approaches to examine valuation-regression models drawn from Berre (2022), which include revenue, country-risk premium (capturing country-level risk-free rate), and sector-level CAPM-beta (capturing sector-level risk premium) as discounted cash flow valuation factors alongside business model.

In particular, revenues can be expected to have positive β-coefficients, while DCF-discount factor components (country-risk premium and CAPM-beta) can both be expected to have negative coefficients. Meanwhile, business model is a categorical variable, which may take the value "business-to-business" (B2B),

Table 4 OLS including DCF valuation factors and business models

OLS coefficients	Estimate	Std. error	T-value	P-value
(Intercept)	5.10E + 08	6.12E + 07	8.326	5.02E − 16***
Revenues	4.27E − 01	6.20E − 02	6.892	1.31E − 11***
Country-risk premium	− 4.19E + 09	3.18E + 09	− 1.317	0.188
Sectoral beta	− 4.61E + 08	6.02E + 07	− 7.664	6.67E − 14***
B2B&C	3.06E + 08	6.13E + 07	4.995	7.59E − 07***
B2B	9.80E + 07	6.25E + 07	1.569	0.117
B2C	6.37E + 08	6.28E + 07	10.138	< 2.00E − 16***

*** $p < 0.01$, ** $p < 0.05$, * $p < 0.1$
Residual standard error: 546,900,000 on 644 degrees of freedom
Multiple R-squared: 0.2793
Adjusted R-squared: 0.2726
F-statistic: 41.6 on 6 and 644 DF, p-value: < 2.20E − 16

Table 5 CART including DCF valuation factors and business model

OBS: 1048				
End nodes: 15				
Complexity parameter	No. of split	RMSE	Cross-validation error	Cross-validation st. dev
0.1280	0	1.0000	1.0024	0.1634
0.0623	2	0.7441	0.8255	0.1484
0.0574	3	0.6817	0.8100	0.1479
0.0376	4	0.6243	0.7245	0.1398
0.0285	5	0.5867	0.7133	0.1397
0.0241	7	0.5296	0.7016	0.1385
0.0148	8	0.5055	0.6458	0.1366
0.0132	9	0.4906	0.6200	0.1317
0.0132	11	0.4643	0.6219	0.1318
0.0102	12	0.4512	0.6242	0.1318
0.0100	13	0.4409	0.6154	0.1318
Variable importance				
Revenue	Business model	Beta	Country-risk premium	
35	24	23	18	

"business-to-customer" (B2C), "business-to-business-and-customers" (B2B&C), or "business-to-government" (B2G).

First, Table 4 examines the relationship between startup valuations, DCF-factors, and business models, splitting business model into dummy variables, using an OLS model, finding that revenue's valuation impact is DCF consistent, while the discount factor appears to be driven by sector-level CAPM-beta. Lastly, the valuation impact of B2B is outweighed by both B2C and B2B&C.

Meanwhile, Table 5 outlines a decision tree-based CART valuation which includes revenue, country-risk premium (capturing country-level risk-free rate), and sector-level CAPM-beta (capturing sector-level risk premium) as discounted cash flow valuation factors alongside business model and describes startup pre-money valuations ranging from €27 million to €3.1 billion, and are partitioned hierarchically.

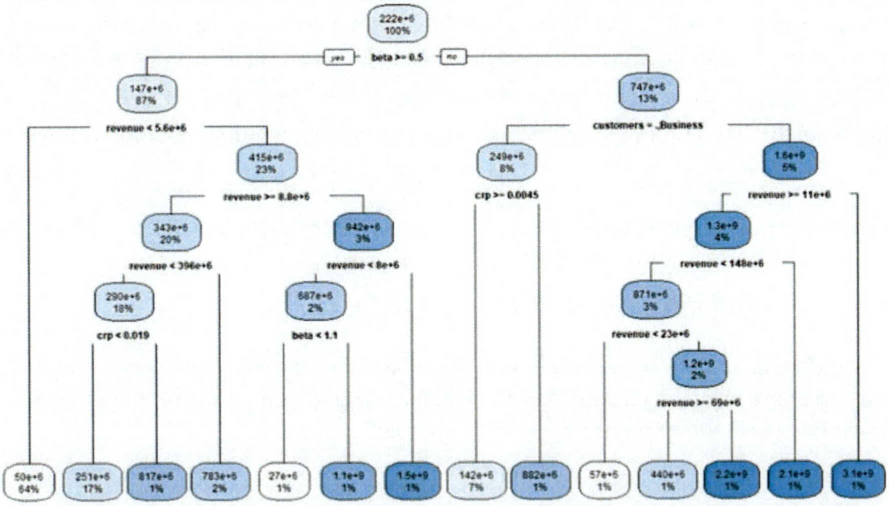

Given the structure of the regression tree in Table 5, the weighted-summation approach and the hierarchical ordinal approach would lead to somewhat-different functional forms. Equation 12 demonstrates a weighted-summation functional form example of the valuation model resulting from Table 4 regression tree, taking the resulting variable-importance indicators as σ-coefficients.

Equation 12 Valuation regression-tree model using weighted-summation segmentation

$$\begin{aligned} \text{Valuation}_i = {} & 0.35\beta_1(\text{Revenue}_i) + 0.24\beta_2(\text{Business Model}_i) \\ & + 0.23\beta_3(\text{Sectoral-Risk Beta}_i) + 0.18\beta_4(\text{Country-Risk Premium}_i) \end{aligned} \tag{12}$$

Using this approach, the highest-valuation tranche would first and foremost consist of startup with substantial revenue figures. This would be followed by firms which have business models, which focus on B2C, B2B&C, or B2G, and whose revenues are discounted by low sector-level CAPM-betas and low country-risk premiums. This means that the highest-valuation EU startup are firms which combine substantial revenue figures with a B2C, B2B&C, or B2G business model, and are located in a low-volatility industry, and based in a AAA-rated home-market such as Germany,

Denmark, or Switzerland (Damodaran, 2021), whereas lowest-valuation EU startup are somewhat more likely to be based in higher-risk European markets (for example in the CEE, Baltic, or Euro-Med regions), and are characterized by high-risk industry sectors, low revenues, and B2B business model. Table 6 presents the regression-tree results outlined in Table 5, as a Payne-Style valuation scorecard.

By also drawing on the OLS findings outlined in Table 2 as a source of β-coefficients, a two-tiered approach is possible. Here, Eq. 13 and Table 6 capture the revisions possible by inclusion of β-coefficients drawn from Table 4. Because business model has been re-transcribed as its constituent (statistically significant) dummy variables, B2C and B2B&C, the functional form of the valuation model includes terms and coefficients for each of these business models, but not B2G nor B2B.

Equation 13 Valuation regression-tree model using weighted-summation segmentation

$$\begin{aligned}\text{Valuation}_i = {} & (0.35 * 0.4273)(\text{Revenue}_i) + (0.24 * 637{,}000{,}000_{\text{B2C}})(\text{Business Model}_i) \\ & + (0.24 * 305{,}900{,}000_{\text{B2B\&C}})(\text{Business Model}_i) \\ & + 0.23\beta_3(-460{,}900{,}000_i) + 0.18\beta_4(._i)\end{aligned} \tag{13}$$

Building on this revision, Table 7 constitutes a revision of the Payne-style summation scorecard, originally outlined in Table 1, featuring incorporation of β-coefficients drawn from use of a two-tiered valuation approach.

Table 6 CART-based valuation as weighted-summation segmentation results presented in Payne-style scorecard

Weighting	Sign of β coef.	Impact on startup valuation
35%	Impact	*Revenue*
	+	Valuation is positively by revenue
		Business model
24%	Impact	*Client focus of the business*
	−	Business-to-business (B2B)
	+	Business-to-customer (B2C)
	+	Business-to-business and customer (B2B&C)
	+	Business-to-government (B2G)
		Discount factor
23%	Impact	*Sector-level CAPM-beta*
	−	Valuation is negatively impacted by sectoral risk
18%	Impact	*Country-risk premium*
	−	Valuation is negatively impacted by country-risk premium
Total		
100%		

Table 7 Two-tiered revised valuation as weighted-summation results presented in Payne-style scorecard

Weighting	β coef.	Impact on startup valuation
35%	Impact	*Revenue*
	0.4273	Valuation is positively by revenue. Per EUR of revenue
		Business model
24%	Impact	*Client focus of the business*
	–	Business-to-business (B2B)—(not significant)
	637,000,000	Business-to-customer (B2C)
	305,900,000	Business-to-business and customer (B2B&C)
	–	Business-to-government (B2G)—(not significant)
		Discount factor
23%	Impact	*Sector-level CAPM-beta*
	– 460,900,000	Valuation is negatively impacted by sectoral risk. Per 1.00 of CAPM-beta
18%	Impact	*Country-risk premium*
	–	Valuation is negatively impacted by country-risk premium. But is not statistically significant within the European EU/EEA dataset. Near-significance indicates that CRP is likely to be significant in more diverse datasets
Total		
100%		

Alternatively, hierarchical ordinal segmentation, the second valuation-segmentation approach, gives rise to a substantially larger and more complex valuation-model functional form, as each of the regression-tree's branch and terminal nodes can be represented in the model. Equation 14 demonstrates an example of the second valuation-segmentation approach, outlined in Eq. 8. Because the CART results include 14 terminal nodes, as well as numerous branch nodes, the size and complexity of the entire long-form valuation equation is substantial.

Equation 14 Valuation regression-tree hierarchical ordinal segmentation model approach

$$\begin{aligned}\text{Valuation}_i = {} & 50{,}000{,}000(\text{Sectoral-Beta} \geq 0.5) * (\text{Revenue}_i < 5{,}600{,}000) \\ & + 251{,}000{,}000(\text{Sectoral-Beta} \geq 0.5) * (\text{Revenue}_i \geq 5{,}600{,}000) \\ & * (\text{Revenue}_i \geq 8{,}800{,}000) * (\text{Revenue}_i < 369{,}000{,}000) \\ & * \left(\text{Country-Risk-Premium}_i < 0.019\right) \\ & + 817{,}000{,}000(\text{Sectoral-Beta} \geq 0.5) * (\text{Revenue}_i \geq 5{,}600{,}000) \\ & * (\text{Revenue}_i \geq 8{,}800{,}000) * (\text{Revenue}_i < 369{,}000{,}000) \\ & * \left(\text{Country-Risk-Premium}_i \leq 0.019\right)\end{aligned}$$

$$
\begin{aligned}
&+ 783{,}000{,}000(\text{Sectoral-Beta} \geq 0.5) * (\text{Revenue}_i \geq 5{,}600{,}000)\\
&* (\text{Revenue}_i \geq 8{,}800{,}000) * (\text{Revenue}_i \geq 369{,}000{,}000)\\
&+ 27{,}000{,}000(\text{Sectoral-Beta} \geq 0.5) * (\text{Revenue}_i \geq 5{,}600{,}000)\\
&* (\text{Revenue}_i < 8{,}800{,}000) * (\text{Revenue}_i < 8{,}000{,}000)\\
&* (\text{Sectoral-Beta} < 1.1)\\
&+ 1{,}100{,}000{,}000(\text{Sectoral-Beta} \geq 0.5) * (\text{Revenue}_i \geq 5{,}600{,}000)\\
&* (\text{Revenue}_i < 8{,}800{,}000) * (\text{Revenue}_i < 8{,}000{,}000)\\
&* (\text{Sectoral-Beta} \geq 1.1)\\
&+ 1{,}500{,}000{,}000(\text{Sectoral-Beta} \geq 0.5) * (\text{Revenue}_i \geq 5{,}600{,}000)\\
&* (\text{Revenue}_i < 8{,}800{,}000) * (\text{Revenue}_i \geq 8{,}000{,}000)\\
&+ 142{,}000{,}000(\text{Sectoral-Beta} < 0.5) * (\text{Business Model}_i = \text{B2B})\\
&* \left(\text{Country-Risk-Premium}_i \geq 0.0045\right)\\
&+ 882{,}000{,}000(\text{Sectoral-Beta} < 0.5)\\
&* (\text{Business Model}_i = \text{B2B}) * \left(\text{Country-Risk-Premium}_i < 0.0045\right)\\
&+ 57{,}000{,}000(\text{Sectoral-Beta} < 0.5)\\
&* (\text{Business Model}_i = \text{B2C or B2B\&C or B2G})\\
&* (\text{Revenue}_i \geq 11{,}000{,}000) * (\text{Revenue}_i < 148{,}000{,}000)\\
&* (\text{Revenue}_i < 23{,}000{,}000)\\
&+ 440{,}000{,}000(\text{Sectoral-Beta} < 0.5)\\
&* (\text{Business Model}_i = \text{B2C or B2B\&C or B2G})\\
&* (\text{Revenue}_i \geq 11{,}000{,}000) * (\text{Revenue}_i < 148{,}000{,}000)\\
&* (\text{Revenue}_i \geq 23{,}000{,}000) * (\text{Revenue}_i \geq 69{,}000{,}000)\\
&+ 2{,}200{,}000{,}000(\text{Sectoral-Beta} < 0.5)\\
&* (\text{Business Model}_i = \text{B2C or B2B\&C or B2G})\\
&* (\text{Revenue}_i \geq 11{,}000{,}000) * (\text{Revenue}_i < 148{,}000{,}000)\\
&* (\text{Revenue}_i \geq 23{,}000{,}000) * (\text{Revenue}_i < 69{,}000{,}000)\\
&+ 2{,}100{,}000{,}000(\text{Sectoral-Beta} < 0.5)\\
&* (\text{Business Model}_i = \text{B2C or B2B\&C or B2G})\\
&* (\text{Revenue}_i \geq 11{,}000{,}000) * (\text{Revenue}_i \geq 148{,}000{,}000)\\
&+ 3{,}100{,}000{,}000(\text{Sectoral-Beta} < 0.5)\\
&* (\text{Business Model}_i = \text{B2C or B2B\&C or B2G})\\
&* (\text{Revenue}_i < 11{,}000{,}000)
\end{aligned}
\tag{14}
$$

An interesting detail about the regression tree, described in Table 4, is that several of the nodes indicate unicorn valuation. Essentially, this tree model appears to contain a recipe for unicorn valuations. Furthermore, we see that revenue drives the majority of the lower and intermediate branches, corroborating revenue's dominant variable-importance role.

Nevertheless, while the entire regression-tree valuation function outlined in Eq. 13 is sizable and cumbersome, it is not necessary to estimate the function as whole. Rather, because segments of the function where the criteria are not met are zero, it suffices to estimate the branches and terminal node where the firm actually finds itself. For example, for a startup located in the rightmost terminal node, whose sectoral beta would be larger than 0.5, and whose revenue is less than €50,000,000, Eq. 15 reduces to

Equation 15 Valuation regression-tree model reduced-form ordinal segmentation model approach

$$\text{Valuation}_i = 50{,}000{,}000(\text{Sectoral-Beta} \geq 0.5) * (\text{Revenue}_i < 5{,}600{,}000) \quad (15)$$

Although this reduced form of the model is both compact and immediately useful for practitioner purposes, substantial detail is lost in terms of other-path branches and terminal nodes, as well as their distributions and threshold values.

6 Discussion and Further Research

Overall, segmented models are historically underappreciated within empirical finance literature, with segmented models surfacing in but a small, obscure fraction of startup-valuation literature (Berre & Le Pendeven, 2022), as well as in startup-selection and startup-survivability models. In particular, opportunities to employ this approach for modelling of startup selection are particularly relevant, given the relative prominence of qualitative decision factors, as outlined in Murnieks et al. (2016) and Andreoli (2022), and as described by Wessendorf et al. (2019).

Nevertheless, appearance of segmented models in industry and practitioner-sourced grey literature (for example, Berkus, 2016; Ernst & Young, 2020; Ewing Marion Kauffman Foundation, 2007; Goldman, 2008; Payne, 2011) serves as an unmistakable indication that segmented approaches have established traction among industry practitioners ranging from business angels and VC investors to auditing and consultancy practitioners.

Why Segmented Models Work?

While these segmented estimation models might presently be under represented within economic literature (and entrepreneurial finance literature in particular), the ongoing proliferation of machine learning techniques can be expected to increase diversity, viability and popularity of segmented models within the literature, given that there are several empirical approaches drawn from both econometrics and machine learning that segmented models can be adapted to. In principle, the industry popularity and usefulness in markets of segmented estimation models can be attributed to numerous noteworthy positive qualities which characterize them.

First, segmented models are directly transposable to empirical modelling, making investigation of their validity and accuracy a relatively straightforward task. Fundamentally, this is the case because both CART and OLS models can be expressed in segmented functional form.

Second, segmented models are mathematically straightforward, making them both straightforward to understand and easy to communicate to clients, investors, and stakeholders. This quality may help explain the widespread popularity of the Berkus and Payne methods among industry practitioners and among industry sources, given that Damodaran (2002) ascribes this quality.

Third, comes their considerable flexibility. Because the segmented estimation-models' functional form are readily transposable for the purposes of empirical modelling, they are also highly adaptable. This means that they can be altered by adding or modifying the impacts of determinant factors as the need arises, for example, by adding segments to capture interaction terms or niche functional form segments. Furthermore, they can be constructed by modifying other styles of selection models, valuation models, survivability models, and stress-testing models. For example, relative-valuation models can be combined into two-factor or three-factor segmented valuation models, while both multi-decision selection models such as Siskos and Zopounidis (1987), and fuzzy analytic hierarchy process outlined in Dhochak and Doliya (2019), can serve as the basis for segmented hierarchical models.

The rise and proliferation of hierarchical empirical approaches, including not only CART-based regression trees, but also related approaches, such as the bottom-up Hierarchical Ascending Classification decision trees, and Random Forest has yielded the proliferation of increasingly accurate and flexible prediction models, which can not only be used for improved accuracy in entrepreneurial finance modelling, but also for speedy decision making, as well as the construction of increasingly flexible segmented models. This indicates that the use of such approaches in the business and market landscape can only be expected to proliferate in future.

Contributions and Further Research

Because this study focuses on the use and import or methodological approaches from industry practitioners, as well as from political science and marketing journals into entrepreneurial finance literature, this study adds to the existing body of research in several ways by both filling existing gaps in the theory, and by elaborating on already existing published empirical findings.

First, this study ties together practitioner approaches and peer-review literature trends. While practitioner-derived or industry-oriented literature such as Ewing Marion Kauffman Foundation (2007) or Ernst and Young (2020) point to segmented valuation models such as those described by Payne (2011) and Berkus (2016), this approach, seen in studies such as Hand (2005) and Sievers et al. (2013) for valuation models and Siskos and Zopounidis (1987) for selection models, is relatively rare within peer-review literature. This may be owed to the overall need for model sophistication in order interaction effects and variable hierarchies within models. This study provides an overview and synthesis of these approaches, which can be generally deployed by practitioners and experts across a wide variety of markets, while also providing context for the ongoing debate within peer-review literature.

Second, this study elaborates on already existing published research in the entrepreneurial finance field. Existing studies which use segmented approaches devote little space to exploring model functional form. Here again, the overall need for model sophistication in order interaction effects and variable hierarchies within models is apparent.

Third, this study describes use of newly emergent empirical techniques and describes how to systematically make use of them in a consistent way. While micro-targeting based on hierarchical decision trees can take several forms in terms of machine learning algorithms (i.e. recursive partitioning, agglomerative hierarchical clustering, Random Forest), the modelling functional form that can be applied for startup valuation, startup selection, or startup survival intended to accompany such modelling approaches has heretofore not yet appeared in literature. This may be owed to the overall novelty of such approaches within entrepreneurial financial literature up until now.

Given that machine learning approaches are generally confronted relatively early on with questions of model selection and algorithm selection, further research using the principles outlined in this paper should consider complexity and shape of functional form as a fundamental part of model selection and algorithm selection, as a combined model outlook. Furthermore, this combined outlook can and should be taken into consideration for all applications of machine learning approaches within finance, economics, or entrepreneurship research, as well and practice thereof in the marketplace.

Implications of this research are far reaching. For markets and industry practitioners, elaboration on why and how hierarchical segmented models work for selection, valuation, and survivability estimation, as well as how they relate to emerging machine learning approaches can lead to the development of new and bespoke entrepreneurial finance models going forward, as industry practitioners may increasing adopt this style of estimation approach. Meanwhile, the emergence of investors linked to the big data and machine learning industries (ranging from CVCs to specialized consultants and experts) may someday try to automate tree-based segmented selection, valuation, and survivability approaches, in contexts where it may be appropriate to do so (for example the implementation of trading bots in a crowdfunding platform or P2P-lending platform setting). For investors, as well as for third parties, implications are also far reaching because these models can hypothetically deliver accurate estimations via microtargeting, which in its purest form is able to bypass difficult to obtain or confidential firm data, making accurate estimations of valuation, selection, and survivability substantially more widespread within startup markets.

Meanwhile, for policy-making circles, implications of proliferation of segmented models as machine learning approaches evolve, develop, and proliferate, might be a more niche and targeting understanding of startup markets, a body of knowledge which may be very useful for the purposes of SME policy, as well as in targeting key sectors, regions, asset classes, or municipalities going forward.

Fundamentally, future research may build on this study by using the principles described here for empirical studies featuring hierarchical machine learning approaches for the development of hierarchical segmented models. Since this approach is still in its emergent phases, it may be feasible to "push-the-envelope" on what is empirically possible. Doing so can be helped, for instance by development of a taxonomy of entrepreneurial finance relevant configurations, clusters, and categorical variables, so that future microtargeting research can grow beyond reliance on industry-sector, business-model, and economic-geography variables (such as cities or postal codes).

Lastly, this research can be used as a roadmap for future studies intending to use either hierarchical machine learning techniques within entrepreneurial finance, for industry practitioners interested in using machine learning techniques to establish bespoke segmented entrepreneurial finance models, or machine learning professionals interested in deploying their expertise for entrepreneurial finance (for example in a fintech setting).

Acknowledgements This article has been sponsored by the "Finance for innovation" Chair at Audencia Business School. Special thanks is accorded to Benjamin Le Pendeven, the chair director.

References

Andreoli, J. J. (2022). *Entrepreneurial finance: Towards determinants for early-stage investments.* ProefschriftMaken. ISBN: 978-90-8980-159-3.

Bellavitis, C., Filatotchev, I., Kamuriwo, D. S., & Vanacker, T. (2017). Entrepreneurial finance: New frontiers of research and practice: Editorial for the special issue embracing entrepreneurial funding innovations. *Venture Capital: An International Journal of Entrepreneurial Finance, 19*(1–2), 1–16.

Benoit, K. (2011). *Linear regression models with logarithmic transformations* [Working Paper]. Methodology Institute London School of Economics.

Berkus, D. (2016). *The Berkus method—Valuing the early-stage investment.* Berkonomics.

Berre, M. (2022). Which factors matter most? Can startup valuation be micro-targeted? In *International Conference on Small Business (ICSB) World Conference* [Working Paper]. Washington, DC.

Berre, M., & Le Pendeven, B. (2022). What do we know about start-up valuation drivers? A systematic literature review. *Venture Capital.* https://doi.org/10.1080/13691066.2022.2086502

Damodaran, A. (2002). *Investment valuation: Tools and techniques for determining the value of any asset.* In Wiley Finance Series. Wiley. ISBN: 9780471414902.

Damodaran, A. (2009). *Valuing young, start-up and growth companies: Estimation issues and valuation challenges* [SSRN Scholarly Paper ID 1418687]. Rochester, NY: Social Science Research Network.

Damodaran, A. (2019). *The disruption dilemma: Valuing the disruptors & disrupted.* NYU Stern. https://pages.stern.nyu.edu/~adamodar/pdfiles/country/Disruption2019.pdf

Damodaran, A. (2021). *Equity risk premiums (ERP): Determinants, estimation, and implications—The 2021 edition*. Available at SSRN: https://ssrn.com/abstract=3825823. https://doi.org/10.2139/ssrn.3825823

Dhochak, M., & Doliya, P. (2019). Valuation of a startup: Moving towards strategic approaches. *Journal of Multi-Criteria Decision Analysis, 27*, 39–49.

Ellis, J. H. M., Williams, D. R., & Roscorla, F. B. (2001). Managing Japan-Europe industrial buyer-supplier relationships: A conceptual and empirical study of the Japanese market for high-technology marine equipment. *Journal of the Asia Pacific Economy, 6*(2), 232–243.

Ernst & Young. (2020). *Startup funding full eGuide—The factory*. EY—The Factory. https://thefactory.works/wp-content/uploads/2020/10/Eguide_Funding_A4.pdf?mc_cid=df85bfb95a&mc_eid=f80b29de26. Accessed June 30, 2022.

Ewing Marion Kauffman Foundation. (2007). *Valuing pre-revenue companies*. Angel Capital Association. https://www.angelcapitalassociation.org/data/Documents/Resources/AngelCapitalEducation/ACEF_-_Valuing_Pre-revenue_Companies.pdf

Fama, E. (1970). Efficient capital markets: A review of theory and empirical work. *The Journal of Finance, 25*, 383–417.

Firmin, S. (2021). *Understanding the outputs of the decision tree tool*. Alteryx Designer Knowledge Base. https://community.alteryx.com/t5/Alteryx-Designer-Knowledge-Base/Understanding-the-Outputs-of-the-Decision-Tree-Tool/ta-p/144773. Accessed September 01, 2022.

Goldman, M. (2008). *Valuation of startup and early-stage companies*. The Value Examiner. http://www.michaelgoldman.com/Publications/Goldman%20Valuation%20of%20Start-ups.pdf

Gonzalez, G. (2017). What factors are causal to survival of a startup? *Muma Business Review, 1*, 097–114.

Hand, J. R. M. (2005). The value relevance of financial statements in the venture capital market. *The Accounting Review, 80*(2), 613–648.

Keene, O. N. (1995). The log transformation is special. *Statistics in Medicine, 14*(8), 811–819.

Khan, I., Capozzoli, A., Corgnati, S. P., & Cerquitelli, T. (2013). Fault detection analysis of building energy consumption using data mining techniques. *Energy Procedia, 42*(2013), 557–566.

Knaup, A. E. (2005). Survival and longevity in the business employment dynamics data. *Monthly Labor Review*, 50–56.

Knaup, A. E., & Piazza, M. C. (2007). Business employment dynamics data: Survival and longevity. *Monthly Labor Review*, 3–10.

Krzywinski, M., & Altman, N. (2017). Classification and regression trees. *Nature Methods, 14*, 757–758. https://doi.org/10.1038/nmeth.4370

Mahmoud, F., Zahoor, A., Hussain, N., & Younes, B. Z. (2022). Working capital financing and firm performance: A machine learning approach. In *Financial Economics Meeting (FEM-2022)*. Paris.

Miller, J., O'Neill, M., & Hyde, N. (2010). *Intermediate algebra* (2nd ed.). McGraw-Hill Higher Education.

Murray, G. R., & Scime, A. (2010). Microtargeting and electorate segmentation: Data mining the American national election studies, *Journal of Political Marketing, 9*(3), 143–166. https://doi.org/10.1080/15377857.2010.497732

Miloud, T., & Cabrol, M. (2011). Les facteurs strategiques inluençant l'evaluation des start-ups par les capitaux-risqueurs. *Revue Management et Avenir, 49*, 36–61.

Murnieks, C. Y., Cardon, M. S., Sudek, R. T., White, D., & Brooks, W. T. (2016). Drawn to the fire: The role of passion, tenacity and inspirational leadership in angel investing. *Journal of Business Venturing, 31*(4), 468–484.

Neter, J., Wasserman, W., & Kutner, M. H. (1990). *Applied linear statistical models: Regression, analysis of variance, and experimental design* (3rd ed.). McGraw-Hill Inc.

Rebonato, R. (2010). *Coherent stress testing: A Bayesian approach to the analysis of financial risk*. In Wiley Finance Series.

Ribeiro-Oliveira, J. P., Garcia de Santana, D., Pereira, V. J., & Machado dos Santos, C. (2018). Data transformation: An underestimated tool by inappropriate use. *Acta Scientiarum Agronomy, 40*(1), e35300. https://doi.org/10.4025/actasciagron.v40i1.35300

Sandeep. (2014). Difference between rel error and xerror in rpart regression trees, version: 2014-07-02. https://stats.stackexchange.com/users/45985/sandeep. URL: https://stats.stackexchange.com/q/105536. Accessed September 01, 2022.

Sievers, S., Mokwa, C. F., & Keienburg, G. (2013). The relevance of financial versus non-financial information for the valuation of venture capital-backed firms. *European Accounting Review, 22*(3), 467–511.

Siskos, J., & Zopounidis, C. (1987). The evaluation criteria of the venture capital investment activity: An interactive assessment. *European Journal of Operational Research, 31*(3), 304–313. http://linkinghub.elsevier.com/retrieve/pii/0377221787900403

Wessendorf, C., Kegelmann, J., & Terzidis, O. (2019). Determinants of early-stage technology venture valuation by business angels and venture capitalists. *International Journal of Entrepreneurial Venturing, 11*(5), 489.

Wooldridge, J. M. (2010). *Econometric analysis of cross section and panel data* (2nd ed.). The MIT Press.

Zopounidis, C., Galariotis, E., Doumpos, M, Sarri, S., & Andriosopoulos, K. (2015). Multiple criteria decision aiding for finance: An updated bibliographic survey, *European Journal of Operational Research, 247*, 339–348.

Zopounidis, C., & Doumpos, M. (2002). Multi-criteria decision aid in financial decision making: Methodologies and literature review. *Journal of Multi-Criteria Decision Analysis, 11*(4–5), 167–186.

Green Versus Non-green Banks: A Differences-In-Differences CAMEL-Based Approach

Ioannis Malandrakis and Konstantinos Drakos

Abstract We employ a panel data set of 165 banks (global and non-global) from thirty-eight countries around the world covering the time period 1999–2015, and we examine whether there are any discernible performance differences between green and non-green banks using panel data techniques (the random effects and the multi-level model). The variables of interest are fundamental CAMEL factors. Moreover, we adopt the Differences-In-Differences approach to examine whether green ("treatment" group) and non-green ("control" group) banks exhibit differential behavior, and we use the outbreak of the financial crisis (2008) as the time of intervention. We find that both green and non-green banks are affected by nearly the same bank-specific factors, and that they do not exhibit heterogeneous behavior with respect to several fundamental aspects. Our results show that green banks perform better than their non-green counterparts only in terms of Total Capital ratio and Tier 1 Capital ratio during and after the financial crisis. As for the rest of the CAMEL factors, it seems that both groups exhibit the same behavior, especially in the post-crisis period. Furthermore, it seems that neither country nor region has any significant effect on CAMEL variables values (it is rather a matter of bank characteristics, either green or non-green). We also find that the financial crisis had (a) a positive effect on capital adequacy (excluding leverage ratio, which seems to have remained unaffected), on asset quality (excluding NPLs ratio) and management quality; (b) a negative effect on earnings ability; and (c) a negative impact on liquidity, for both bank types.

Keywords CAMEL factors · Financial crisis · Green banking · Panel data models · Sustainable finance

JEL Classification C32 · C33 · G01 · G21 · G29 · Q56

I. Malandrakis (✉) · K. Drakos
Department of Accounting and Finance, School of Business, Athens University of Economics and Business, 76, Patission Str., 10434 Athens, Greece
e-mail: imalandrakis@aueb.gr

K. Drakos
e-mail: kdrakos@aueb.gr

C. Zopounidis et al. (eds.), *Operational Research Methods in Business, Finance and Economics*, Lecture Notes in Operations Research,
https://doi.org/10.1007/978-3-031-31241-0_3

1 Introduction

Global warming is the main element of climate change and is mainly attributed to the extended use of fossil fuels and CO_2 emissions. IPCC (2021, p. 18) mentions that an increase of the average global temperature by 2 °C is quite possible (if not imminent) between 2041 and 2060 in the intermediate scenario. Figure 1 provides a simple but convincing evidence on the close relationship between CO_2 emissions and global temperature change.

The severity of climate change is reflected into climate-driven natural disasters effect on humans and the economy; Eckstein et al. (2020) mention that, between 1999 and 2018, more than 495,000 people died as a direct result of more than 12,000 extreme weather events, while the economic losses at the same time period were approximately 3.54 trillion USD (in Purchasing Power Parities). Figure 2 depicts global temperature anomalies (temperature change) and natural disasters occurrence from 1880 to 2017 (note the co-movement of global temp. change and natural disasters increase after the mid-70s).

Despite the fact that less-developed countries are generally more affected than developed and industrialized countries, in recent years advanced economies start to feel the climate change impact more clearly than ever before. To this extent, climate change-driven natural disasters affect countries' economies and financial sectors. Natural disasters amplify the fragility of banking systems as they increase banks'

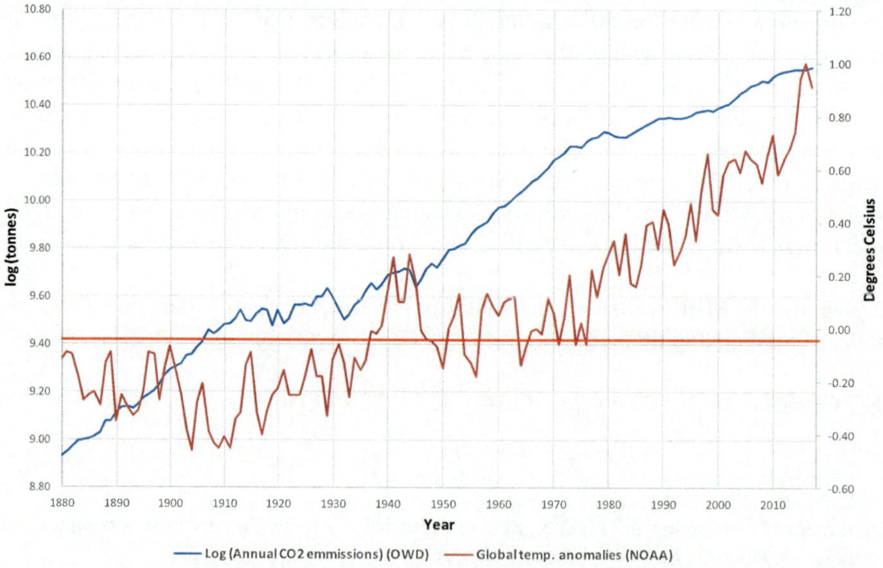

Fig. 1 CO_2 emissions and global temperature: 1880–2017 *Data source* Annual CO_2 emissions: our world in data; global temperature anomalies: National Oceanic and Atmospheric Administration (NOAA)

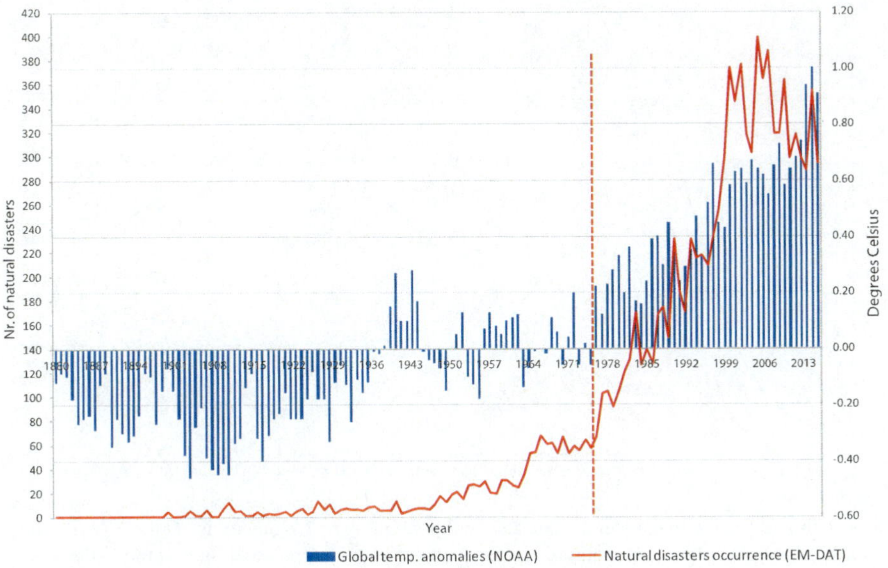

Fig. 2 Global temperature anomalies and natural disasters occurrence 1880–2017 *Data source* Global temperature anomalies: NOAA. Natural disasters: EM-DAT

probability of default (Danmarks Nationalbank, 2019; Klomp, 2014), while Lamperti et al. (2019) show that climate change could increase banking crisis probability from 26 to 248% by 2100 (under the extreme scenario).

The need for the transition from a high- to a low-carbon economy has led to the creation of (a) green banking (credit facilities to low-carbon investments, renewables, etc.), (b) green banks (fully public and quasi-public), and (c) pure green banks which are dedicated public or non-profit finance entities focusing mainly on increasing and accelerating investments in clean power and services using finance tools to mitigate climate change (CGC, 2019).

Banks can have a significant contribution to protect the environment and help reduce the negative impacts of climate change by (a) modifying their lending lines ("green credit lines", "green loans") and policies, (b) adjusting their risk management to reflect climate considerations, and (c) incorporating climate change risk and sustainability issues into their lending decisions and long-term strategic thinking (Boston Common Asset Management, 2015; IFC-World Bank, 2010). Moreover, banks could have potential economic benefits from environment protection and their becoming greener, as they can benefit in terms of (a) lower cost of capital from improved environmental risk management, (b) improved quality of their loan portfolio (i.e., lower NPLs), (c) development of profitable business lines by integrating environmental and sustainability issues into their lending decisions, (d) better ratings by analysts, and (e) new business opportunities in green technologies and underserved markets (see, e.g., IFC-World Bank, 2010; Sharfman & Fernando, 2008).

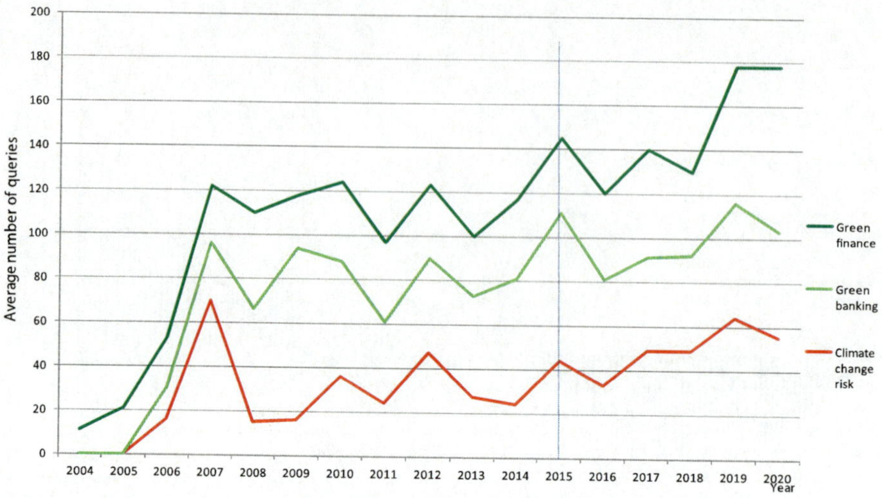

Fig. 3 Climate change risk, green banks, and banking: Google trends analytics *Data source* https://trends.google.com; values per year are simple averages of the corresponding monthly values

Moreover, the interest in green banking, green finance, and climate change risk is steadily increasing in recent years as shown in Fig. 3.

In this study, we compare green and non-green banks in order to examine if green banks differ from their non-green counterparts in terms of the CAMEL ratios (solvency, asset quality, management quality, profitability, and liquidity). We control for the performance of green versus non-green banks before and after the financial crisis using a Differences-In-Differences (DID) approach, and we apply a random effect as well as a multilevel (or hierarchical) model to control for country and/or region effects. For the purposes of our analysis, we employ an array of variables using a panel data set of 165 banks from 38 countries worldwide for the period 1999–2015, and we classify banks as green and non-green on the basis of certain criteria (see Sect. 3, Subsect. 3.2.1). We also subclassify our sample data set as global and non-global banks, because global banks are exceptionally important for their investment and lending activities worldwide but also suspect for contributing to the transmission of financial crises across banking systems and economies (see, e.g., Cetorelli & Goldberg, 2010; Hale et al., 2016; Kalemli-Ozcan et al., 2013; Moosa, 2010).

To the best of our knowledge, no other empirical study exists comparing green and non-green banks in terms of CAMEL/S ratios using a data set of both global and non-global banks from around the world.

The remainder of this paper is organized as follows: Sect. 2 presents related literature. Section 3 describes data, variables, and methodology. Section 4 presents the results (*t*-test, correlation analysis, unit root test, and regressions) and discusses random effects and multilevel regressions results. Finally, Sect. 5 concludes and presents policy implications.

2 Related Literature

2.1 CAMEL/CAMELS Rating System

CAMEL/CAMELS rating system is a widely accepted supervisory tool for assessing a bank's overall condition (FDIC, 2018). Its bank-specific variables are well documented since they have been used in numerous banking sector studies (see, e.g., Anastasiou et al., 2019; Beck et al., 2013; Berger & DeYoung, 1997; Chiaramonte et al., 2015; Christopoulos et al., 2011; Cole & White, 2017; Doumpos & Zopounidis, 2010; Gilbert et al., 2000; Gopalan, 2010; Papanikolaou, 2018; Swindle, 1995; Van den End, 2013; Whalen, 2005). After the financial crisis, special emphasis has been placed on the leverage ratio as a basic capital measure alongside the risk-based capital ratios (see BCBS, 2019). CAMEL/CAMELS rating and variables are also used in measuring banks' and banking systems risk and financial health (see, e.g., Chernykh & Kotomin, 2022; Peria Martinez & Schmukler, 2001),[1] banks' performance from a supervisory point of view (Hirtle et al., 2020; Kupiec et al., 2017)[2] as well as in comparative studies of different banking systems (see, e.g., Afroj, 2022; Nguyen et al., 2020)[3] and in constructing new indexes for measuring banks' financial strength (Doumpos et al., 2017).[4]

---

[1] They measure Russian banking system risk using, among other ratios, capital adequacy ratios, and CAMEL/S ratings as determinants of Deposit Insurance premia, and they consider that the cost of insured deposit remains a predictor of bank failures beyond the CAMEL variables in Russia or in other risk-based deposit insurance (RBDI) schemes banking systems.

[2] Kupiec et al. (2017) examine the impact of poor bank supervisory CAMEL/S rating on banks' loan growth. They use CAMEL ratings 1–5 plus the CAMEL variables leverage capital ratio, past due to assets, liquid assets to total assets, ROA before tax, and log of real assets (a size proxy). Hirtle et al. (2020) use some CAMEL variables (size, loans/assets%, NPLs%, ROA%, Tier 1%) and ratings (1–5); they find that top-ranked banks (e.g., those with better-valued CAMEL variables and higher CAMEL ratings (1 = best rating, 5 = worst rating)) that receive more supervisory attention hold less risky loan portfolios, are less volatile, and are less sensitive to industry downturns, but do not have lower growth or profitability.

[3] Afroj (2022) studies the financial strength of the Bangladesh banking sector using CAMEL ratios and finds that Islamic banks are more robust—in terms of capital adequacy and liquidity—and financially stronger—in comparison with the conventional and the Islamic window banks. Nguyen et al. (2020) examine the effects of CAMEL variables on the financial performance of Vietnamese banks and show that capital adequacy, asset quality, management efficiency, and liquidity strongly affect the financial performance (measured in terms of ROA, ROE, and net interest margin) of banks in Vietnam.

[4] Doumpos et al. (2017) compare Islamic versus conventional banks using a data set of Islamic and conventional banks from 57 countries (members of the Organisation of Islamic Cooperation) ending in a data set of 101 Islamic banks, 347 conventional banks, and 52 banks with an Islamic banking window operating in 21 countries over the period 2000–2011; they employ, among other variables, traditional financial ratios associated with the CAMEL rating system embodied into a single overall financial strength indicator, namely the *Bank Overall Financial Strength Index (BOFSI)* and point out, among other things, the usefulness of aggregating traditional financial ratios associated with the CAMEL rating system into a single overall financial strength indicator that can form the basis of a monitoring system.

2.2 Green or Sustainable Finance

According to UNEP Inquiry (2016), there is no single or globally accepted definition for "green" or "sustainable finance", while terms like "green finance", "sustainable finance", "climate finance", and "low-carbon finance" are often used interchangeably. Sustainable finance has a broader meaning as it includes jointly social, environmental, and economic aspects for the financial services sector and aims at supporting sustainable economic growth (EBA, 2019), while green finance excludes social aspects and takes into account environmental issues and risks and focuses on the flow of private funds (but not only) toward green investments, i.e., into the green industries or sectors of the economy. G20 Green Finance Study Group (2016) defines as green (or sustainable) finance "[...] *the financing of investments that provide environmental benefits (e.g. reductions in CHG emissions), covering a wide range of financial institutions and asset classes and involving the effective management of environmental risks across the financial system*". Green finance can take the form of "green loans" and "green bonds" (for financing low-carbon activities and as a necessary tool for large long-term infrastructure investments—Sartzetakis, 2019).

2.3 Green Banking and Green Banks

Green banking could be defined in terms of (a) banking commitments (implementation of green finance principles), (b) financial flows (the volume and distribution of banks' loans to green investments), (c) financial risk (e.g., the impact on NPLs and ROE), and (d) environmental and social outcomes (avoidance of negative E&S impacts, etc.) (IFC-World Bank & SBN, 2017). Again, there is no uniform or globally accepted definition for green banks, but there are definitions for the following:

- "Pure green banks": These are dedicated public or non-profit finance entities focusing mainly on increasing and accelerating investments in clean power and services using finance tools to mitigate climate change (CGC, 2019).
- "Green Investment Banks" (GIBs): These are public or quasi-public institutions which finance renewable energy, energy efficiency, and other clean energy projects in partnership with private lenders aiming also to advance public objectives (CGC, 2017; Gilleo et al., 2016; OECD, 2015).
- "Social banks": They consider not only profit but also environment and people; in some cases, the concepts of green and social bank coincide (Benedikter, 2011).

2.4 *International Organizations, Green Banking, and Sustainable Finance*

There is a number of international non-governmental organizations (NGOs), non-profit institutions, and initiatives that join together financial institutions from all over the world under a common goal: to protect the environment by promoting a new lending framework.

The *United Nations Environment Programme—Finance Initiative* (UNEP FI) is a partnership between the United Nations and the global financial sector created in the wake of the 1992 Earth Summit with a mission to promote sustainable finance (UNEP FI, 2021a). Currently, it has more than 400 financial institutions as members, including banks, insurers, and investors; as of September 2021, UNEP FI had 307 bank members (May 2019: 140 bank members) (UNEP FI, 2021b).

The *Equator Principles* (EP) was formed in 2003 and is a risk management framework, adopted by financial institutions, for determining, assessing, and managing environmental and social risk in projects' finance (EP, 2021a). By the end of September 2021, 126 financial institutions from 38 countries worldwide[5] have adopted Equator Principles (EP, 2021b).

The *Global Alliance for Banking on Values* (GABV) was founded in 2009 and is a network of banking leaders from around the world aiming to change the banking system so that it become more transparent and to support economic, social, and environmental sustainability (GABV, 2021a). As of August 2021, the GABV had 70 member banks (GABV, 2021b).

The *Banking Environment Initiative* (BEI) was founded in 2010 and has as basic mission to direct capital toward environmentally friendly projects. Its operation is supported by the University of Cambridge Institute for Sustainability Leadership (CISL, 2018).

The *Ceres* (or *CERES*) is a sustainability non-profit organization which was founded in 1989, and it is a national coalition of investors, environmental groups, and other public interest organizations that work together with companies to address sustainability challenges such as climate change. CERES also directs the investor network on climate risk, a group of 60 institutional investors from the USA and Europe managing over $4 trillion of assets (Ceres, 2021; Cogan, 2008).

The *U.S. Alliance for Sustainable Finance* (USASF) was formed in December 2018, is based in New York City, and its basic aim is to drive investments' financing in clean energy and climate resilience projects across the USA, supporting the reduction of GHG emissions. The founding members are 15 leading banks (such as Bank of America, Citi, Credit Suisse, Wells Fargo) and investment companies (U.S. Alliance for Sustainable Finance, 2019).

[5] From 70 in 2010 (IFC-World Bank, 2010).

On the regulatory and supervision side, there is the *NGFS*[6] and the *European Commission Initiative on Sustainable Finance.* This EC initiative involves a devoted section on sustainable finance through the setup of a technical expert group (TEG) in December 2016 and a high-level expert group (HLEG) in December 2018 engaged in sustainable finance, special reports drafting, legislative proposals, and high impact conferences. A key document is the "Action Plan on Financing Sustainable Growth" (released in March 2018) which sets an EU strategy on sustainable finance and a roadmap for future work—through ten actions—across the financial system (European Commission, 2018).

3 Data, Variables and Methodology

3.1 Data

We use data from Thomson Reuters Eikon (ex-DataStream) for the CAMEL factors, and from (a) supranational organizations (Federal Stability Board, Basel Committee on Banking Supervision) as well as from the "Banks Around the World" website and (b) non-governmental organizations and institutions (BEI, EP, GABV, and UNEP FI) for the classification of banks as global and green, respectively. In addition, to confirm if a bank is green or non-green as well as global or not, according to the developed classification criteria (see Sect. 3, Subsect. 3.2.1), we used bank-specific info from banks' annual reports and/or sustainability reports through their official websites. From our initial sample, observations in which specific CAMEL variables are above 1.5 times the 99.5th percentile or below 0.5 times the 0.5th percentile are characterized as outliers and excluded from the analysis. Then observations in which specific CAMEL ratios values are above the 99.5th percentile or below the 0.5th percentile are also removed. Our final—unbalanced panel data set comprises 2805 observations in total, from 1999 to 2015[7] on a yearly basis, and of specific CAMEL variables for 165 banks from 38 countries worldwide (see Fig. 4).

---

[6] In December 2017 at the Paris "One Planet Summit", eight central banks and supervisors from around the world established the Central Banks and Supervisors Network for Greening the Financial System (NGFS) which has a basic aim in contributing to the best possible extent in the achievement of the "well below 2° Celsius' goal" that was set out in the Paris agreement and promoting environmental sustainable growth in line with financial stability goals (NGFS, 2018, 2019).

[7] Each CAMEL bank-specific variable is as of December 31 of each successive data year.

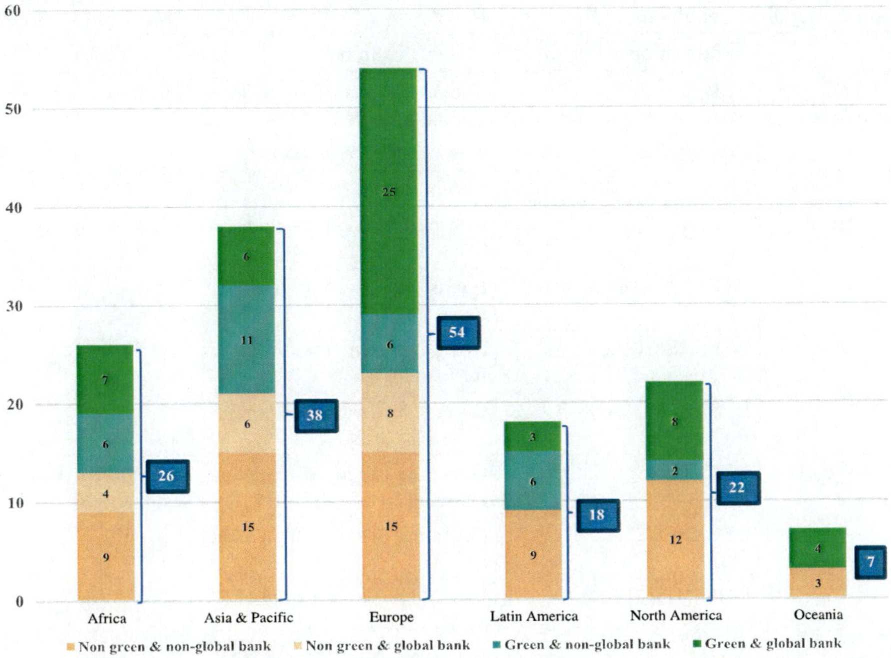

Fig. 4 Green and non-green (global and non-global) banks per region (as of December 31, 2015)

3.2 *Variables*

We employ fourteen quantitative bank-specific variables, which are thirteen CAMEL factors plus bank size (see Table 1). Moreover, we use three dummy variables which stand for the characterization of a bank as global (or not) and as green (or not), for country, for financial crisis period, and for the interaction between crisis and green banks (see Table 2). As for the crisis variable, we set as cutoff point for the financial crisis outburst the last quarter of 2007; specifically, we define as pre-crisis and post-crisis period the time period before 2007 Q4 and after 2007 Q4, respectively. These cutoff dates are based on the timelines outlined by the Bank for International Settlements (BIS) (2009). The data set is not split into more sub-periods, i.e., in pre-crisis, crisis, and post-crisis phase, given that the duration of these phases and their effects on banks presumably was not uniform worldwide: Crisis length was certainly different for South European Countries (e.g., Greece, Portugal, Spain, Italy), compared to the other European countries (e.g., Germany, France, Norway, etc.), the USA, etc. (see, e.g., Baur, 2012; Basten & Serrano, 2019; De Bondt & Vermeulen, 2021; Laeven & Valencia, 2012). Note that the expressions "post-crisis" and "during and after crisis" are used interchangeably, denoting essentially the same time period

Table 1 Variables definitions

	Abbreviation/acronym	Variable definition	Data source
Capital adequacy	TCR	Capital adequacy ratio or Total Capital ratio (%)	DataStream
	CRTIER1	Tier 1 Capital ratio (%)	
	LR	Leverage ratio (%)	
Asset quality	NPLS	Non-performing loans/total loans (%) (NPLs)	DataStream
	PROVLLLOANS	Provision for loan losses/total loans (%)	
	NPLSRESERLL	Non-performing loans/reserve for loan losses (%)	
	RESERLLLOANS	Reserve for loan losses/total loans (%)	
Management quality	OPEXPENSOPINCOME	Operating expenses/operating income (%)	DataStream; own estimations
	OPEXPENSTA	Operating expenses/total assets (%)	
	NONINTEXPENSTA	Non-interest expenses/total assets (%)	
Earning ability	ROA	Return on assets (%) (ROA)	DataStream
	ROE	Return on equity (%) (ROE)	
Liquidity	LTD	Total loans/total deposits (%) (L-t-D)	DataStream
	SIZE	Bank size (= log of total assets)	DataStream; own estimations

Notes (1) Operating expenses, operating income, non-interest expenses, and total assets amounts (all in 000's USD) are from Thomson Reuters Eikon (ex-DataStream) database, while *operating expenses/operating income (%), operating expenses/total assets (%), non-interest expenses/total assets (%),* and *bank size* variables are own estimations based on DataStream's relevant data. (2) Total loans = loans − interbank loans

(i.e., from January 2008 to December 2015) although the expression "during and after crisis" refers to the time segment 2008–2009, which represents the peak of the global financial crisis, and it is used as the appropriate time point for the interaction of green bank with crisis.

3.2.1 Classification Criteria

First, we classify a bank as "global" if it (a) operates in more than one continent (or region) via subsidiaries or branches including banks with significant number of representative offices around the world and (b) has an average asset size equal or more to 100 billion USD during the whole sample data period; the first criterion is a

Table 2 Dummy variables definition

Abbreviation/ acronym	Variable	Variable definition	Data source
GLOBAL	Global	0 if non-global and 1 if global	FSB, BCBS, "Banks Around the World" (http://www.relbanks.com), own research
NEWGREEN	Green	0 if non-green and 1 if green	BEI, EP, GABV, UNEP FI, own research
CRISIS	Crisis	0 if time ≤ 2007 Q4 and 1 if time ≥ 2008 Q1	

Notes (1) The characterization of a bank as global or not is based not only on available data sources but also to own research according to the developed classification criteria (see Sect. 3, Subsect. 3.2.1). (2) The assignment of the value 1 for green and 0 for non-green bank is in accordance to the developed classification criteria (see Sect. 3, Subsect. 3.2.1). (3) Cutoff dates for financial crisis period (pre and post-crisis) are set according to the BIS (2009) timelines

necessary one, while the second is not.[8] These criteria are in line with BCBS (2013, 2017), Fillat et al. (2013), De Haas and Van Lelyveld (2014), and Jeucken (2001, p. 186) definitions. Also, the majority of the banks that are characterized as global are included in the FSB (2016) list of G-SIBs and in the "Banks Around the World" website[9] for 2015–2016. In addition, all banks classified as global or multinationals were confirmed as such through investigation of their official websites (e.g., annual report, bank presentation).

Next, we classify a bank as "green" if it was a member, by the end of 2015, at least in one of the following non-governmental and supranational organizations: *Banking Environment Initiative* (BEI), *Equator Principles* (EP), *United Nations Environment Programme—Finance Initiative* (UNEP FI), and *Global Alliance on Banking on Values* (GABV) (see Sect. 2, Subsect. 2.4). This is consistent with Benedikter (2011, pp. 43–45) analysis for social and green banks. Additional criteria (e.g., the soundness of the applied sustainability policy for environment's protection, the types of green loans offered, green loans' amounts, and their share in banks' loan portfolio, etc.) are not employed because such data were not available. However, to confirm that a certain bank is indeed green according to the above-mentioned classification criterion, an individual search was performed by reviewing banks' annual sustainability reports covering the vast majority of the banks characterized as green. Definitions of variables and relevant data sources are depicted in Tables 1 and 2.

[8] Note however that the concepts of global bank and G-SIB are not always coinciding; it is possible for a global bank not to be a G-SIB if it does not meet all of the necessary criteria. Also the terms "multinational" and "global" seem to have equal meaning (see for instance De Hass & Van Lelyveld, 2014; Niepmann, 2011), although "global bank" has presumably a broader meaning than "multinational bank": A multinational bank can operate in more than one countries within the same continent/region (e.g., Europe) while a global bank can operate in more than one continents excluding parent institution's continent/region (for instance a European multinational bank in Latin America, Africa, etc.).

[9] http://www.relbanks.com.

Finally, according to our classification criteria, out of 165 banks, 84 are identified as green (of which 53 global and 31 non-global) and 81 as non-green (of which 18 global and 63 non-global). Figure 4 summarizes green (either global or not) and non-green (either global or not) banks region, as of December 31, 2015. Table 3 presents green and non-green banks per continent and country. Special care was taken during sample selection, so that green banks, per region and country, do not exceed significantly in number non-green banks and vice versa.

3.3 Methodology

Our analysis involves three stages:

(a) Hypothesis testing (using the *T*-test approach) to examine whether there are statistically significant differences in the mean values of CAMEL factors between green and non-green banks, both in the pre- (prior to 2008) and post-crisis period (2008 onwards) as well as in the whole sample data period (1999–2015). The associated test hypotheses are

   1. H_0: mean $CAMEL_{\text{i green bank}}$ = mean $CAMEL_{\text{i non-green bank}}$
   2. H_0: mean $CAMEL_{\text{i all banks pre-crisis}}$ = mean $CAMEL_{\text{i all banks post-crisis}}$
   3. H_0: mean $CAMEL_{\text{i green banks pre-crisis}}$ = mean $CAMEL_{\text{i green banks post-crisis}}$
   4. H_0: mean $CAMEL_{\text{i non-green banks pre-crisis}}$ = mean $CAMEL_{\text{i non-green banks post-crisis}}$
   5. H_0: mean $CAMEL_{\text{i green banks pre-crisis}}$ = mean $CAMEL_{\text{i non-green banks pre-crisis}}$
   6. H_0: mean $CAMEL_{\text{i green banks post-crisis}}$ = mean $CAMEL_{\text{i non-green banks post-crisis}}$.

   Rejecting all or some of the above hypotheses would imply that the average values of each variable for green versus non-green banks, during the whole data period as well as in the pre- and post-crisis period, have differences, and thus, there exists a primary indication that green and non-green banks exhibit statistically significant behavior with respect to all or some of their CAMEL factors between them during each examination period.

(b) Unit root test and correlation analysis.

(c) Use of panel data regression techniques to investigate whether there are statistically significant differences between the two basic groups allowing for the appropriate dynamics to capture potential persistence. Our analysis takes into account bank country of origin, fixed time, and bank effects. Specifically, we estimate a series of models, using both fixed and random effects, ending in a DID approach to address whether green and non-green banks exhibit differential behavior. Following the general framework for Differences-In-Differences estimation described by Imbens and Wooldridge (2007) and the methodology employed for the banking sector outlined by Berger and Roman (2020), we define green banks as the "treatment" group and non-green banks as the "control" group, and we use the financial crisis outbreak as the time of "intervention".

Table 3 Green banks per region and country

Region	Country	Banks (total)	*Of which*: green banks	%	*Of which*: global banks	%
Africa	Kenya	2	1	50	0	0
	Mauritius	2	1	50	0	0
	Morocco	2	1	50	0	0
	Nigeria	12	6	50	3	50
	South Africa	8	4	50	4	100
	Total	26	13	50	7	54
Asia and Pacific	Bangladesh	2	1	50	0	0
	China	8	4	50	2	50
	India	2	1	50	0	0
	Indonesia	2	1	50	0	0
	Japan	10	4	40	3	75
	Malaysia	2	0	0	0	–
	Oman	2	1	50	0	0
	South Korea	6	3	50	1	33
	Taiwan R.O.C.	2	1	50	0	0
	Thailand	2	1	50	0	0
	Total	38	17	45	6	35
Europe	Belgium	2	1	50	0	0
	Denmark	2	1	50	0	0
	France	6	4	67	4	100
	Germany	4	1	25	0	0
	Greece	2	0	0	0	–
	Italy	4	2	50	2	100
	Netherlands	3	2	67	2	100
	Norway	2	1	50	1	100
	Spain	8	5	63	5	100
	Sweden	4	4	100	4	100
	Switzerland	6	2	33	2	100
	Turkey	6	3	50	0	0
	UK	5	5	100	5	100
	Total	54	31	57	25	81
Latin America	Argentina	2	1	50	0	0
	Brazil	6	3	50	3	100
	Chile	2	1	50	0	0
	Colombia	2	1	50	0	0

(continued)

Table 3 (continued)

Region	Country	Banks (total)	*Of which*: green banks	%	*Of which*: global banks	%
	Ecuador	2	1	50	0	0
	Mexico	2	1	50	0	0
	Peru	2	1	50	0	0
	Total	18	9	50	3	33
North America	Canada	10	5	50	5	100
	USA	12	5	42	3	60
	Total	22	10	45	8	80
Oceania	Australia	7	4	57	4	100
	Total	7	4	57	4	100
	Grand total	165	84	51	53	63

Notes (1) Banks are classified as *global, green* according to the developed classification criteria (see Sect. 3, Subsect. 3.2.1). (2) Data sources: for green banks: BEI, EP, UNEP FI, GABV, and own research; for global banks: Financial Stability Board (FSB), "Banks Around the World" (http://www.relbanks.com), and own research

Additional analysis was performed, employing multilevel modeling to control for country and region effects.

3.3.1 Panel Data Regression Models, Main Hypotheses, and Variables

The general form of the panel data regression model can be written as (see Eq. 1):

$$Y_{it} = a + \beta X_{it} + u_{it}, \tag{1}$$

where i denotes entity and t time, α is a scalar, β is $K \times 1$, X_{it} is the itth observation on K explanatory variables, and u_{it} is the error term (Baltagi, 2005, p. 11).

Our panel data regression model in its combined form can be written as follows (see Eq. 2):

$$Y_{ijt} = a + \beta X_{ijt-1} + \gamma \mathrm{DV}_s + u_{it} + \varepsilon_{it}, \tag{2}$$

where i denotes entity (bank), j country, and t time (year from 1999 to 2015), Y_{ijt} is the dependent variable accounting for CAMEL variables for bank i in country j in year t, α is the unknown intercept for each entity i being estimated using both fixed (FE) and random effects (RE), X_{ijt-1} is a vector of lagged bank-level control variables (CAMEL factors, bank size), and DV_s is a vector of other control variables (including our key or main variables) expressed as dummy variables, and u_{it} and ε_{it} are the between entity error and the within error, respectively.

Furthermore, we choose to cluster on the bank rather than country level as some of the countries in our sample have significantly more banks than others (see Fig. 4 and Table 3).

We estimate our model by adding successively bank-level control variables, and the dummy variables *global bank*, *country*, *crisis*, and *green bank* and finally an *interaction term* to capture the effect of crisis (time of intervention) on green banks (treatment group). Equation (3) represents the fully expanded version of our model:

$$Y_{ijt} = a + \beta_i X_{ijt-1} + \gamma_1 \mathrm{GL}_{ijt} + \gamma_2 C_j + \gamma_3 \mathrm{CR}_t + \gamma_4 \mathrm{GR}_{ijt} + \gamma_5 \mathrm{CR}_t \mathrm{GR}_{ijt} + u_{it} + \varepsilon_{it}, \quad (3)$$

where i denotes entity (bank), j country, and t time (year from 1999 to 2015), Y_{ijt} is the dependent variable accounting for CAMEL variables for bank i in country j in year t, α is the unknown intercept for each entity i being estimated using both fixed (FE) and random effects (RE), X_{ijt-1} is a vector of lagged bank-level control variables (CAMEL factors, bank size), GL_{ijt} is a dummy variable accounting for global bank taking the value one for global bank and zero otherwise, C_j stands for country dummy, CR_t is a dummy variable which takes the value zero before and one after crisis, GR_{ijt} is a dummy variable taking the value one if a bank is classified as green bank and zero otherwise, $\mathrm{CR}_t \times \mathrm{GR}_{ijt}$ is the interaction term which captures the effect of crisis on bank type (green and non-green), and u_{it} and ε_{it} are the between entity error and the within error, respectively.

By adding $\mathrm{CR}_t \times \mathrm{GR}_{ijt}$, we specifically test for significant differences between green and non-green banks during and after the global financial crisis. Equation (3) is the fully expanded version of our model and essentially represents the DID-type approach applied to assess differences between green and non-green banks with respect to CAMEL factors, where non-green bank is the control group, green bank is the treatment group, and the financial crisis outbreak (after 2007 Q4) is the time of intervention.

Despite the fact that we estimated our model using both FE and RE models, we believe that the RE model is the most appropriate form[10] since it allows for time invariant variables[11] such as country of origin and global bank, while in the FE model such variables' effects are absorbed by the constant term. Moreover, if there are indications that the differences across entities (i.e., banks) may affect the dependent variable/-s, then the RE model is presumably the most appropriate form (Clark & Linzer, 2015).

[10] Although the *Hausman test* proposes in many cases FE versus RE model, the choice between the two types is not as easy as it might seem (see Baltagi, 2005, pp. 18–19), especially if we take into account that this test does not always provide a clear result and in most cases, it favors FE against RE.

[11] See Wooldridge (2002, p. 288).

Furthermore, we test three main hypotheses:

(a) H_0: $\gamma_3 = 0$, i.e., crisis has not affected CAMEL factors, against the alternative Ha: $\gamma_3 \neq 0$, i.e., crisis has affected CAMEL factors, concerning both banks' types; if CR estimate is not statistically significant, we accept H_0 and reject Ha.
(b) H_0: $\gamma_4 = 0$, i.e., bank type (green or non-green) does not affect CAMEL factors, against the alternative Ha: $\gamma_4 \neq 0$, i.e., bank type affects CAMEL factors (e.g., green banks differ from non-green banks in terms of Total Capital ratio and so on); if GR estimate is not statistically significant, we accept H_0 and reject Ha.
(c) H_0: $\gamma_5 = 0$, i.e., crisis has not affected bank type (green or non-green) with respect to CAMEL factors, against the alternative Ha: $\gamma_5 \neq 0$, i.e., crisis has affected bank type (green or non-green) with respect to CAMEL factors (for instance green banks are better that non-green banks after crisis in terms of Total Capital ratio and so on); if CR × GR estimate is not statistically significant, we accept H_0 and reject Ha.

The logic behind our hypotheses is that we seek to explore if there are differences among the two bank types—on the basis of their risk profile and financial health as reflected by CAMEL variables—before and especially after the 2007–2008 financial crisis (which is considered as the major event, i.e., the time of intervention for both green and non-green banks either global or not); if any differences are found, then which is the direction of these differences? Moreover, we formulated the above-mentioned hypotheses (and especially hypotheses (b) and (c)) by assuming that green banks could have a hypothetically superior performance than non-green banks by assuming a better loan portfolio (e.g., a lower NPLs ratio), a higher liquidity ratio (e.g., a lower Loan-To-Deposit ratio) and so forth, and consequently better CAMEL ratios as a result of the gradually increasing number of green loans.[12] These loans may be less risky than normal loans, due to the anticipated lower risk weighting to such exposures because of the so-called green-supporting factor (GSF) (see, e.g., Dunz et al., 2021) as well as of the simultaneous implementation of the GSF and the "brown-penalizing factor" (BPF) or "dirty penalizing factor" (DPF) (see, e.g., Dafermos & Nikolaidi, 2021).

The final number of variables employed (after correlation analysis) is sixteen, specifically (a) thirteen control variables which are ten CAMEL variables, bank size, and two dummy variables which account for global bank (GL) and countries (C), (b) two key variables that are the dummy variables which stand for crisis (CR) and green bank (GR), and (c) the effect of crisis on green banks (CR × GR), which is also a key variable.

With respect to the last variable, we try to capture the effect of financial crisis on green banks through the use of an additional dummy variable, i.e., of an interaction term defined as *crisis × green bank*; more specifically, we interact our basic key variable of interest, that is the green bank dummy, with the crisis dummy variable, and we focus on this interaction term which is a moment in time where the fact that a

[12] Which are special terms loans with lower interest rates that make green lending in general more attractive for the banking sector (see Subsect. 2.2).

bank is green (or not) can become critical. In this case, γ_5, i.e., the coefficient of the interaction variable (see Eq. 3), captures the impact of financial crisis (CR) on the dependent variable Y_i, when GR equals one (and vice versa), not the impact of crisis on Y in general; moreover, we have a two-way interaction which describes the effect of a joint increase of CR and GR on Y_i (see, e.g., Brambor et al., 2006; Braumoeller, 2004).

An additional analysis was performed using a multilevel (or hierarchical) model. In addition to the RE model (see Eqs. 1–3), we employ a multilevel model[13] to explore for country and region effects regarding our variables of interest. A multilevel model can be used whenever data could be nested in more than one category, i.e., in countries and regions. By applying such a model, we can study effects that vary by groups (e.g., regions) (see, e.g., Kayo & Kimura, 2011; Makridou et al., 2019).[14] Our observations or data points per bank (i.e., all CAMEL variables of interest plus dummy variables and the interaction term) are grouped or nested at three levels: at bank, country, and at region level; that is obs. per bank is the lowest level, country is the next level (level 2), and region is the top level (level 3).[15] Moreover, given that we are interested in green banks, we additionally define *country* × *green bank* as the second level and region *country* × *green bank* as the third level. Thus, we consider that observations across banks are nested in a given country within a region and that banks (green) are cross-classified with countries and regions.

Following Gelman and Hill (2007, pp. 1–2), the general form of the employed multilevel regression model can be written as (see Eqs. 4 and 5):

$$Y_i = a_{ji} + \beta X_i + \varepsilon_i, \text{ for bank } i = 1, 2, \ldots, n \tag{4}$$

$$a_j = a + bu_j + \eta_j, \text{ for country } j = 1, 2, \ldots, J, \tag{5}$$

where i stands for the individual bank and $j[i]$ for the country j containing bank i; α generally represents the overall intercept and can be considered as the grand mean (e.g., of the TCR, and so forth).

---

13 Multilevel models are called hierarchical for two different reasons (Gelman & Hill, 2007, p. 2): first, from the structure of the data (in our case: banks clustered within countries); and second, from the model itself, which has its own hierarchy, with the parameters of the within countries regressions at the bottom, controlled by the hyperparameters of the upper-level model (i.e., at the region level in our case). Multilevel models are also known as mixed-effect models that include both fixed and random effects (Gelman & Hill, 2007, p. 2).

14 Kayo and Kimura (2011) analyze the direct and indirect effects of firm/industry/country characteristics on firms leverage. Makridou et al. (2019) examine, among other things, the effect of time, firm, and country characteristics on the financial performance (profitability) of the firms participating in the EU emissions trading scheme.

15 Through a multilevel modeling approach, we can assess the link between the external environment (i.e., country, region) and the internal characteristics of the banks (i.e., green banks), distinguishing between bank-level variability and variability across countries and regions.

Note that, X_i and u_j represent predictors at the bank and country levels, respectively, and ε_i and η_j are independent error terms at each of the two levels.[16] A third level (i.e., the top level) can be added following the notation of Eq. (5). Finally, our complete multilevel model takes the following form:

$$Y_{ijk} = a + \beta X_{(t-1)ijk} + \gamma \text{DV}_s + c_{ij} + r_{ik} + \varepsilon_{tijk}, \tag{6}$$

where a is the overall intercept (or grand mean), $X_{(t-1)ijk}$ is a vector of CAMEL variables, DV_s is a vector of dummy variables stand for global bank, crisis, green bank, and interaction crisis × green bank (see Eq. 3), c_j and r_k, correspond to the random effects representing the country and region level, respectively; ε_{tijk} is the random error term that stands for the variance across *i*, *j*, *k*, time.

An alternative version of Eq. (6) that incorporates interactions between green banks and countries and regions is specified as follows:

$$Y_{ijk} = a + \beta X_{(t-1)ijk} + \gamma \text{DV}_s + ic_{ij} + ir_{ik} + \varepsilon_{tijk}, \tag{7}$$

where a is the overall intercept (or grand mean), $X_{(t-1)ijk}$ is a vector of CAMEL variables, γDV_s is a vector of dummy variables stand for global bank, crisis, green bank, and interaction crisis × green bank (see Eq. 3), ic_{ij} and ir_{ik} correspond to the random effects representing the interaction between country and green bank, and region and green bank, respectively; ε_{tijk} is the random error that stands for the variance across *i*, *j*, *k*, time.

To sum up, we employ a three-level multilevel model applying a maximum likelihood estimation (MLE) procedure, considering that observations per bank per year are nested within a given country, and countries are nested within regions.

4 Empirical Analysis

Descriptive statistics are given in Table 4. Unit root test results are shown in Table 5. *T*-test results are presented in Table 6. Correlation matrices for the variables under investigation are depicted in Table 7. In Table 8, the estimation results of the fully expanded version (see Eq. 3) of the random effects (RE) model are presented,[17] while the estimation results of the fully expanded version (see Eqs. 6 and 7) of the multilevel are depicted in Tables 9 and 10.

---

[16] The number of "data points" J (countries) in the next-level regression is typically much less than n, the sample size of the lower-level model (banks).

[17] Due to space limitations, the results of all RE model specifications 1–8 are not reported here. We have also employed the fixed effects (FE) model. Both models' results are available from the authors upon request.

Table 4 Descriptive statistics

Variables	All banks			(1) Green versus non-green banks: whole period						(2) All banks (green and non-green): pre- versus post-crisis period					
				Green banks only			Non-green banks only			Pre-crisis period only			Post-crisis period only		
	Obs	Mean	Standard deviation	Obs	Mean	Standard deviation	Obs	Mean	Standard deviation	Obs	Mean	Standard deviation	Obs	Mean	Standard deviation
TCR	1601	14.35	4.98	700	14.38	3.86	901	14.34	5.70	563	12.68	5.12	1038	15.26	4.66
CRTIER1	1626	11.59	4.95	723	11.17	3.47	903	11.93	5.86	592	9.98	5.49	1034	12.51	4.36
LR	2214	18.81	13.16	792	17.95	9.20	1422	19.28	14.90	1050	18.51	14.48	1164	19.07	11.83
NPLS	1955	3.40	5.89	742	2.71	3.05	1213	3.82	7.05	895	2.71	4.14	1060	3.99	6.98
PROVLLLOANS	2320	1.23	2.63	842	0.95	1.38	1478	1.40	3.11	1093	1.15	2.33	1227	1.31	2.87
NPLSRESERLL	1677	116.24	66.75	657	124.20	67.01	1020	111.10	66.10	766	98.51	60.25	911	131.14	68.32
RESERLLLOANS	2220	3.24	4.35	808	2.48	2.09	1412	3.68	5.17	1044	3.38	4.14	1176	3.11	4.52
OPEXPENSOPINCOME	2094	536.11	392.09	774	556.41	372.51	1320	524.20	402.79	1013	556.68	395.14	1081	516.83	388.40
OPEXPENSTA	2334	6.72	7.02	847	5.51	3.31	1487	7.42	8.35	1099	7.68	8.88	1235	5.88	4.62
NONINTEXPENSTA	2300	3.14	3.56	847	2.73	2.02	1473	3.37	4.17	1086	3.44	4.48	1214	2.87	2.44
ROA	2064	1.94	5.64	750	1.64	4.79	1314	2.12	6.02	970	2.01	4.87	1094	1.88	6.20
ROE	2337	11.16	32.66	857	12.37	24.97	1480	10.45	36.37	1087	14.33	30.19	1250	8.39	34.45
LTD	2288	117.66	57.16	827	121.97	56.30	1461	115.22	57.52	1078	118.76	58.62	1210	116.69	55.83
SIZE	2456	17.88	2.18	878	19.14	1.96	1578	17.18	1.98	1156	17.55	2.18	1300	18.18	2.15

(continued)

Table 4 (continued)

Variables	(3) and (5) Green banks pre- versus post-crisis (green banks only)						(4) and (6) Non-green banks pre- versus post-crisis (non-green banks only)					
	Pre-crisis period (green banks only)			Post-crisis period (green banks only)			Pre-crisis period (non-green banks only)			Post-crisis period (non-green banks only)		
	Obs	Mean	Standard deviation	Obs	Mean	Standard deviation	Obs	Mean	Standard deviation	Obs	Mean	Standard deviation
TCR	208	12.08	2.22	492	15.35	4.00	355	13.02	6.20	546	15.19	5.18
CRTIER1	231	8.91	2.40	492	12.23	3.39	361	10.66	6.68	542	12.77	5.07
LR	275	16.55	7.74	517	18.69	9.81	775	19.21	16.16	647	19.37	13.23
NPLS	261	1.52	1.57	481	3.36	3.44	634	3.20	4.73	579	4.51	8.88
PROVLLLOANS	291	0.67	1.35	551	1.10	1.38	802	1.33	2.57	676	1.48	3.65
NPLSRESERLL	227	92.88	58.15	430	140.74	65.51	539	100.88	61.01	481	122.56	69.68
RESERLLLOANS	277	2.13	1.82	531	2.66	2.20	767	3.83	4.63	645	3.49	5.75
OPEXPENSOPINCOME	281	548.04	311.98	493	561.18	403.19	732	559.99	422.89	588	479.64	371.84
OPEXPENSTA	293	5.81	2.97	554	5.35	3.47	806	8.36	10.13	681	6.30	5.35
NONINTEXPENSTA	284	2.76	2.01	543	2.71	2.03	802	3.68	5.05	671	3.00	2.73
ROA	264	1.38	0.94	486	1.78	5.91	706	2.25	5.66	608	1.97	6.42
ROE	299	15.19	39.88	558	10.86	10.04	788	14.01	25.60	692	6.40	45.33
LTD	289	130.04	60.95	538	117.64	53.19	789	114.63	57.23	672	115.92	57.89
SIZE	302	19.11	1.86	576	19.16	2.02	854	17.00	2.02	724	17.40	1.92

Notes All variables are defined in Table 1. The sample data period is between December 1999 and December 2015 (yearly observations). All data are available from Thomson Reuters Eikon (ex-DataStream); OPEXPENSOPINCOME, OPEXPENSTA, NONINTEXPENSTA, and SIZE variables are own estimations based on Thomson Reuters Eikon (ex-DataStream) relevant data

Table 5 Unit root test

Variable	Fisher-type test: PP		Fisher-type test: ADF	
	Inverse χ^2 P		Inverse χ^2 P	
	Statistic	*p*-value	Statistic	*p*-value
TCR	580.2269	0.0000	548.2267	0.0000
CRTIER1	483.0542	0.0000	404.5226	0.0000
LR	732.1591	0.0000	676.2576	0.0000
NPLS	661.8412	0.0000	908.6079	0.0000
PROVLLLOANS	1242.8755	0.0000	1198.1790	0.0000
NPLSRESERLL	442.9218	0.0000	413.0789	0.0000
RESERLLLOANS	891.6500	0.0000	1058.6913	0.0000
OPEXPENSOPINCOME	1455.9198	0.0000	1822.8309	0.0000
OPEXPENSTA	1050.2379	0.0000	928.2495	0.0000
NONINTEXPENSTA	833.3788	0.0000	719.2283	0.0000
ROA	–	–	–	–
ROE	1404.0071	0.0000	1332.9615	0.0000
LTD	809.8842	0.0000	603.6942	0.0000
SIZE	376.8398	0.0384	742.0795	0.0000

Notes (1) Fisher-type unit root test refers to the *Phillips–Peron* (*PP*) and *augmented Dickey–Fuller* (*ADF*) tests, respectively; PP test was performed with constant term, lag(1) and ADF test with constant term, lag(0) and no trend. (2) H_0: all panels contain unit roots; Ha: at least one panel is stationary. (3) All series—variables are stationary

4.1 Descriptive Statistics

Table 4 presents descriptive statistics for bank-specific CAMEL factors and bank size as follows: for all banks (green and non-green), for green versus non-green banks during the whole sample data period (panel 1), for all banks in the pre- and the post-crisis period (panel 2), for green banks in the pre- and the post-crisis period (panels 3 and 5), and non-green banks in the pre- and the post-crisis period (panels 4 and 6).

4.2 Unit Root Test

We examine our variables for unit roots by employing the *Phillips–Peron Fisher-type* test and the *augmented Dickey–Fuller* (*ADF*) test (Choi, 2001). The null hypothesis (H_0) is that all panels contain a unit root against the alternative (Ha) that at least one panel is stationary.

The results reject the null hypothesis for the existence of a unit root in our panel data set, i.e., all series variables are stationary.

Table 6 Hypothesis testing: *t*-test results

	(1) Green versus non-green banks: whole period		(2) All banks: pre- versus post-crisis period		(3) Pre- versus post-crisis (green banks only)		(4) Pre- versus post-crisis (non-green banks only)		(5) Green versus non-green banks: only pre-crisis period		(6) Green versus non-green banks: only post-crisis period	
Variable	Obs	Unequal variances	Obs	Unequal variances	Obs	Unequal variances	Obs	Unequal variances	Obs	Unequal variances	Obs	Unequal variances
TCR	1601	– 0.1803 (0.8570)	1601	– 9.9647*** (0.0000)	700	– 13.8043*** (0.0000)	901	– 5.4503*** (0.0000)	563	2.5987*** (0.0096)	1038	– 0.5680 (0.5702)
CRTIER1	1626	3.2554*** (0.0012)	1626	– 9.6283*** (0.0000)	723	15.1051*** (0.0000)	903	– 5.0966*** (0.0000)	592	4.5566*** (0.0000)	1034	2.0383** (0.0418)
LR	2214	2.6056*** (0.0092)	2214	– 0.9847 (0.3249)	792	– 3.3635*** (0.0008)	1422	– 0.2113 (0.8327)	1050	3.5655*** (0.0004)	1164	1.0108 (0.3123)
NPLS	1955	4.7927*** (0.0000)	1955	– 5.0137*** (0.0000)	742	– 9.9829*** (0.0000)	1213	– 3.1623*** (0.0016)	895	7.9354*** (0.0000)	1060	2.8553*** (0.0044)
PROVLLLOANS	2320	4.7722*** (0.0000)	2320	– 1.4483 (0.1477)	842	– 4.3973*** (0.0000)	1478	– 0.9010 (0.3678)	1093	5.5052*** (0.0000)	1227	2.4969** (0.0127)
NPLSRESERLL	1677	– 3.9293*** (0.0001)	1677	– 10.3913*** (0.0000)	657	– 9.5962*** (0.0000)	1020	– 5.2576*** (0.0000)	766	1.7137* (0.0873)	911	– 4.0586*** (0.0001)
RESERLLLOANS	2220	7.6799*** (0.0000)	2220	1.4597 (0.1445)	808	– 3.6542*** (0.0003)	1412	1.2337 (0.2175)	1044	8.5428*** (0.0000)	1176	3.3706*** (0.0008)

(continued)

Table 6 (continued)

	(1) Green versus non-green banks: whole period		(2) All banks: pre- versus post-crisis period		(3) Pre- versus post-crisis (green banks only)		(4) Pre- versus post-crisis (non-green banks only)		(5) Green versus non-green banks: only pre-crisis period		(6) Green versus non-green banks: only post-crisis period	
Variable	Obs	Unequal variances	Obs	Unequal variances	Obs	Unequal variances	Obs	Unequal variances	Obs	Unequal variances	Obs	Unequal variances
OPEXPENSOPINCOME	2094	– 1.8531* (0.0640)	2094	2.3252** (0.0202)	774	– 0.5053 (0.6135)	1320	3.6695*** (0.0003)	1013	0.4915 (0.6232)	1081	– 3.4308*** (0.0006)
OPEXPENSTA	2334	7.8066*** (0.0000)	2334	6.0408*** (0.0000)	847	1.9991** (0.0460)	1487	4.9967*** (0.0000)	1099	6.4341*** (0.0000)	1235	3.7731*** (0.0002)
NONINTEXPENSTA	2300	4.9755*** (0.0000)	2300	3.6995*** (0.0002)	827	0.3844 (0.7008)	1473	3.2392*** (0.0012)	1086	4.2515*** (0.0000)	1214	2.1799** (0.0295)
ROA	2064	2.0019** (0.0454)	2064	0.5267 (0.5985)	750	– 1.4458 (0.1488)	1314	0.8294 (0.4070)	970	3.9387*** (0.0001)	1094	0.5179 (0.6047)
ROE	2337	– 1.5062 (0.1322)	2337	4.4426*** (0.0000)	857	1.8462* (0.0658)	1480	3.9007*** (0.0001)	1087	– 0.4762 (0.6342)	1250	– 2.5110** (0.0122)
LTD	2288	– 2.7335*** (0.0063)	2288	0.8647 (0.3873)	827	2.9147*** (0.0037)	1461	– 0.4278 (0.6688)	1078	– 3.7380*** (0.0002)	1210	– 0.5362 (0.5919)

Notes (1) Absolute values in parentheses denote p-values. (2) Number of asterisks denotes significance level: *** p-value < 0.01 (statistically significant at the 1% level), ** 0.01 < p-value < 0.05 (statistically significant at the 5% level), * 0.05 < p-value < 0.1 (statistically significant at the 10% level). (3) t-test critical values: $t = 1.96$, $p = 0.05$: if $t \geq 1.96$ and $p < 0.05$, there is a statistical significant difference, otherwise no

Table 7 Correlation matrix 2: final variables

Variable	TCR	CRTIER1	LR	NPLS	PROVLL LOANS	NPLS RESERLL	OPEXPENS OPINCOME	ROA	ROE	LTD	SIZE
TCR	1.0000										
CRTIER1	**0.9455**	1.0000									
LR	0.2381	0.2279	1.0000								
NPLS	0.1984	0.2457	0.0627	1.0000							
PROVLLLOANS	0.5563	0.5066	0.2187	0.3838	1.0000						
NPLSRESERLL	0.0613	0.0973	– 0.0668	0.4905	– 0.0468	1.000					
OPEXPENSOPINCOME	– 0.0732	– 0.0653	0.0479	0.1481	0.1134	0.1618	1.0000				
ROA	0.4590	0.4526	0.3179	– 0.0554	0.4034	– 0.2324	– 0.2228	1.0000			
ROE	0.0244	0.0268	– 0.1158	– 0.2140	– 0.2162	– 0.1407	– 0.1019	0.3969	1.0000		
LTD	– 0.1084	– 0.1048	0.4884	0.0302	– 0.0011	0.0577	0.2053	0.0845	0.0347	1.0000	
SIZE	– 0.3183	– 0.3855	– 0.1653	– 0.1177	– 0.1969	– 0.0862	– 0.0110	– 0.3969	– 0.0298	0.0097	1.0000

Notes (1) All variables are defined in Table 1. (2) Data sources: Thomson Reuters Eikon (ex-DataStream) and own estimations. (3) Pairwise correlations above $r = 0.65$ are marked with bold characters

Table 8 RE model results: fully expanded version (Eq. 3/specification 9)

Independent variables	Dependent variable									
	TCR [1]	CRTIER1 [2]	LR [3]	NPLS [4]	PROVLL LOANS [5]	NPLS RESERLL [6]	OPEXPENS OPINCOME [7]	ROA [8]	ROE [9]	LTD [10]
$TCR_{(T-1)}$	0.584*** (6.15)		0.034 (0.80)	0.061 (0.50)	– 0.295*** (2.88)	– 0.453 (0.88)	– 14.164*** (3.20)	0.016** (2.03)	0.043 (0.32)	0.011 (0.04)
$CRTIER1_{(T-1)}$		0.596*** (5.50)								
$LR_{(T-1)}$	0.025 (1.64)	0.017 (1.10)	0.813*** (5.50)	– 0.019 (0.84)	0.003 (0.63)	– 0.270 (0.83)	0.422 (0.15)	0.0004 (0.07)	– 0.260 (1.36)	0.380* (1.68)
$NPLS_{(T-1)}$	0.041 (0.64)	0.046 (0.86)	– 0.068 (1.13)	0.783*** (13.98)	0.013 (0.77)	0.944 (1.29)	21.747 (1.54)	– 0.016 (0.88)	– 0.813 (1.72)	– 0.408 (1.08)
$PROVLLLOANS_{(T-1)}$	0.121 (1.00)	0.050 (0.45)	0.083 (0.49)	– 0.134 (0.58)	0.677*** (12.23)	– 2.621* (1.85)	– 4.971 (0.34)	0.074*** (2.61)	1.271 (1.62)	0.961 (1.00)
$NPLSRESERLL_{(T-1)}$	– 0.002 (0.84)	– 0.0001 (0.08)	– 0.002 (0.77)	– 0.001 (0.68)	0.0002 (0.35)	0.745*** (21.35)	– 0.456 (1.69)	– 0.001 (1.65)	– 0.019 (1.12)	0.009 (0.54)
$OPEXPENSOPINCOME_{(T-1)}$	– 0.00001 (0.05)	– 0.0002 (0.99)	0.0003 (0.55)	0.0005* (1.98)	0.0004*** (2.73)	0.007* (1.81)	0.354 (5.80)	– 0.0003** (2.55)	– 0.013* (1.87)	– 0.002 (0.67)
$ROA_{(T-1)}$	– 0.035 (0.20)	– 0.046 (0.24)	– 0.198 (0.70)	0.059 (0.39)	0.321*** (2.78)	5.601*** (2.78)	23.493 (1.64)	0.527*** (884)	0.177 (0.18)	– 1.879 (1.36)
$ROE_{(T-1)}$	0.0027 (1.02)	– 0.0005** (2.19)	0.035*** (4.04)	– 0.011** (2.47)	0.008*** (3.47)	– 0.138*** (2.90)	– 0.731 (0.93)	– 0.006 (1.44)	0.010 (0.25)	0.067* (1.92)

(continued)

Table 8 (continued)

Independent variables	Dependent variable									
	TCR [1]	CRTIER1 [2]	LR [3]	NPLS [4]	PROVLL LOANS [5]	NPLS RESERLL [6]	OPEXPENS OPINCOME [7]	ROA [8]	ROE [9]	LTD [10]
$LTD_{(T-1)}$	– 0.007*** (3.40)	– 0.007** (2.19)	– 0.002 (0.33)	0.002 (0.82)	– 0.0003 (0.39)	0.082* (1.95)	0.700*** (2.35)	0.001 (0.93)	0.016 (0.57)	0.779*** (11.66)
$SIZE_{(T-1)}$	– 0.178 (1.68)	– 0.205 (1.51)	0.129 (1.27)	0.019 (0.24)	0.028 (0.73)	1.344 (1.02)	– 30.568** (2.04)	– 0.002 (0.06)	– 1.125 (0.67)	– 0.874 (1.66)
GLOBAL	0.224 (0.88)	0.646*** (2.79)	– 0.493 (1.33)	0.400 (1.27)	– 0.259** (2.01)	– 6.872* (1.71)	11.096 (0.23)	0.135 (0.92)	4.204 (0.77)	2.663* (1.88)
COUNTRY	YES	YES	YES	YES	YES	YES	YES	YES	YES	YES
CRISIS	**1.308*** (3.72)**	**1.340*** (3.25)**	0.031 (0.09)	**0.791*** (2.77)**	**0.243*** (2.86)**	**14.434*** (2.74)**	**70.037** (2.02)**	**– 0.306*** (3.11)**	**– 7.032*** (2.96)**	– 2.309 (1.35)
NEWGREEN	– 0.165 (0.65)	**– 0.495** (2.02)**	– 0.236 (0.69)	– 0.341 (1.40)	0.051 (0.81)	– 1.916 (0.43)	– 20.262 (0.55)	– 0.002 (0.03)	0.957 (0.56)	– 0.404 (0.20)
INTERACTION (= CRISIS × NEWGREEN)	**0.568** (2.26)**	**0.596** (2.28)**	0.394 (0.64)	– 0.043 (0.16)	– 0.097 (1.11)	8.074 (1.48)	63.911 (1.60)	0.027 (0.22)	3.541 (0395)	– 1.118 (0.48)
CONSTANT	8.742*** (2.98)	8.154** (2.50)	1.322 (0.64)	– 0.914 (0.51)	– 0.719 (0.91)	– 2.515 (0.09)	902.886*** (2.96)	0.730 (0.92)	44.327 (1.23)	30.373** (2.36)
Observations	901	927	889	897	916	869	856	874	911	915
No. of countries	31	32	31	31	31	31	30	31	31	31

(continued)

Table 8 (continued)

Independent variables	Dependent variable									
	TCR [1]	CRTIER1 [2]	LR [3]	NPLS [4]	PROVLL LOANS [5]	NPLS RESERLL [6]	OPEXPENS OPINCOME [7]	ROA [8]	ROE [9]	LTD [10]
Diagnostics										
R^2 (overall)	0.782	0.813	0.840	0.450	0.745	0.751	0.480	0.707	0.131	0.879
Wald chi2(1) test: *p*-value	0.0000	0.0000	0.0000	0.0000	0.0000	0.0000	0.0000	0.0000	0.0000	0.0000

Notes (1) Method of estimation: RE generalized least squares (GLS). (2) Group variable: bank. (3) All independent variables are defined in Tables 1 and 2. (4) Absolute values in parentheses denote heteroskedasticity-robust *z-statistics* (two-tailed test): *** statistically significant at the 1% level, ** statistically significant at the 5% level, and * statistically significant at the 10% level. (5) In columns, 1, 3–6 and 8–10, seven countries were omitted because of collinearity; in column 2, six countries were omitted because of collinearity; in column 7, eight countries were omitted because of collinearity. (6) The number of observations varies across regressions, depending on the missing values for the bank-specific variables

Bold indicate significance at the 1% and 5% for the specific variables

Table 9 Multilevel model results: fully expanded version (Eq. 6/specification 9)

Independent variables	Dependent variables									
	TCR [1]	CRTIER1 [2]	LR [3]	NPLS [4]	PROVLL LOANS [5]	NPLS RESERLL [6]	OPEXPENS OPINCOME [7]	ROA [8]	ROE [9]	LTD [10]
$TCR_{(T-1)}$	0.625*** (27.36)		0.007 (0.18)	0.073** (2.28)	− 0.029*** (3.48)	− 0.129 (0.35)	− 7.669*** (2.84)	0.012 (1.69)	0.017 (0.08)	− 0.184 (0.96)
$CRTIER1_{(T-1)}$		0.636*** (30.01)								
$LR_{(T-1)}$	0.182 (0.013)	0.011 (0.86)	0.865*** (38.33)	− 0.015 (0.81)	0.003 (0.67)	− 0.255 (1.25)	− 1.489 (0.99)	0.003 (0.74)	− 0.302** (2.48)	0.418*** (3.87)
$NPLS_{(T-1)}$	0.139 (0.384)	0.037 (0.95)	− 0.026 (0.39)	0.783*** (14.36)	0.003 (0.22)	0.544 (0.84)	13.096** (2.70)	− 0.011 (0.98)	− 0.628 (1.60)	− 0.118 (0.37)
$PROVLLLOANS_{(T-1)}$	0.069 (0.081)	− 0.021 (0.28)	0.035 (0.25)	− 0.145 (1.27)	0.743*** (25.05)	− 2.574* (2.00)	11.809 (1.23)	0.061** (2.50)	1.492 (1.86)	1.154 (1.71)
$NPLSRESERLL_{(T-1)}$	− 0.001 (0.57)	− 0.00002 (0.01)	− 0.003 (1.00)	− 0.0006 (0.28)	0.0005 (0.82)	0.790*** (30.41)	− 0.344 (1.82)	− 0.001** (2.35)	− 0.031 (1.92)	− 0.002 (0.13)
$OPEXPENSOPINCOME_{(T-1)}$	0.00005 (0.19)	− 0.0002 (0.87)	0.0002 (0.46)	0.0007* (1.98)	0.0005*** (5.34)	0.007 (1.78)	0.500*** (14.85)	− 0.0003*** (3.72)	− 0.013*** (5.08)	− 0.002 (0.95)
$ROA_{(T-1)}$	− 0.255 (0.22)	− 0.020 (0.17)	− 0.184 (0.91)	0.060 (0.36)	0.273*** (6.43)	4.414** (2.32)	− 2.241 (0.16)	0.688*** (18.33)	0.979 (0.81)	1.420 (1.46)

(continued)

Table 9 (continued)

Independent variables	Dependent variables									
	TCR [1]	CRTIER1 [2]	LR [3]	NPLS [4]	PROVLL LOANS [5]	NPLS RESERLL [6]	OPEXPENS OPINCOME [7]	ROA [8]	ROE [9]	LTD [10]
$ROE_{(T-1)}$	0.001 (0.22)	– 0.007 (1.64)	0.034*** (4.71)	– 0.010 (1.76)	0.010*** (6.17)	– 0.127 (1.84)	– 1.300 (0.97)	– 0.007 (1.88)	0.019 (0.40)	0.067 (1.87)
$LTD_{(T-1)}$	– 0.006*** (3.05)	– 0.005*** (2.79)	0.003 (0.92)	0.0009 (0.32)	– 0.0006 (0.76)	0.067* (2.08)	0.557** (2.33)	0.001** (2.10)	0.117 (0.58)	0.831*** (45.84)
$SIZE_{(T-1)}$	– 0.208*** (– 3.23)	– 0.218*** (3.47)	0.015 (0.14)	0.052 (0.57)	0.042 (1.80)	2.071* (2.01)	– 5.291 (0.69)	– 0.036 (1.79)	– 1.620** (2.53)	– 0.897 (1.66)
GLOBAL	0.447** (2.00)	0.642*** (2.98)	– 0.312 (0.81)	0.470 (1.48)	– 0.042** (2.38)	– 4.483 (1.25)	2.480 (0.10)	0.650 (0.96)	5.181** (2.30)	1.601 (0.86)
COUNTRY	–	–	–	–	–	–	–	–	–	–
CRISIS	– 1.312*** (5.96)	1.296*** (6.18)	0.179 (0.05)	0.752** (2.38)	0.237*** (2.88)	10.595*** (2.88)	32.359 (1.21)	– 0.246*** (3.51)	– 5.578** (– 2.28)	– 2.751 (1.47)
NEWGREEN	– 0.075 (0.28)	– 0.364 (1.46)	– 0.108 (0.23)	– 0.390 (1.00)	0.021 (0.21)	– 3.344 (0.74)	– 35.087 (1.10)	0.039 (0.47)	0.853 (0.29)	– 0.368 (0.16)
INTERACTION (= CRISIS × NEWGREEN)	0.471 (1.56)	**0.535 (1.90)***	0.432 (0.81)	– 0.166 (0.38)	– 0.103 (0.91)	7.050 (1.38)	39.419 (1.08)	0.042 (0.43)	3.928 (1.14)	0.214 (0.08)
CONSTANT	8.500*** (6.74)	7.730*** (6.39)	2.187 (1.00)	– 1.676 (0.94)	– 0.991** (2.16)	– 25.048 (1.22)	408.846*** (2.68)	1.275*** (3.15)	54.206*** (4.18)	34.397*** (3.20)

(continued)

Table 9 (continued)

Independent variables	Dependent variables									
	TCR [1]	CRTIER1 [2]	LR [3]	NPLS [4]	PROVLL LOANS [5]	NPLS RESERLL [6]	OPEXPENS OPINCOME [7]	ROA [8]	ROE [9]	LTD [10]
Random effects parameters										
Region level	0.005 (0.117)	0.000 (0.000)	0.000 (0.000)	0.030 (0.132)	0.000 (0.000)	10.668 (16.864)	1695.864 (1543.521)	0.064 (0.044)	0.000 (0.000)	12.426 (10.896)
Country level	0.486 (0.236)	0.064 (0.260)	1.041 (0.595)	0.548 (0.333)	0.033 (0.019)	34.432 (22.594)	2069.465 (1869.55)	0.005 (0.012)	0.000 (0.000)	12.614 (12.077)
Residuals	4.283 (0.206)	3.896 (0.185)	13.137 (0.650)	9.006 (0.435)	0.625 (0.030)	1192.876 (58.096)	60,363.300 (3006.442)	0.435 (0.218)	583.389 (27.338)	322.526 (15.658)
ICC/variance decomposition										
Region level %	0.00	0.00	0.00	0.00	0.00	0.01	0.08	2.12	0.00	0.15
Country level %	1.27	0.03	0.62	0.337	0.28	0.08	0.12	0.01	0.00	0.15
Individual level %	98.73	99.97	99.38	99.63	99.72	99.91	99.80	97.87	100.00	99.70
Diagnostics										
Log likelihood	– 1953.09	– 1969.54	– 2421.81	– 2272.34	– 1097.10	– 4320.12	– 5937.88	– 887.47	– 4193.67	– 3.954.52
Wald chi2(1) test: *p*-value	0.0000	0.0000	0.0000	0.0000	0.0000	0.0000	0.0000	0.0000	0.0000	0.0000
Observations	901	927	889	897	916	869	856	874	911	915
No. of groups										
Regions	6	6	6	6	6	6	6	6	6	6
Countries	32	33	32	32	32	32	31	32	32	32

Notes (1) Method of estimation: MLE. (2) Nesting levels at: region, country level. (3) Constant estimate represents the intercept estimate and can be seen as the grand mean corresponding to each dependent variable [1]–[10]. (4) All independent variables are defined in Tables 1 and 2. (5) Absolute values in parentheses denote *z-statistics* (two-tailed test): *** statistically significant at the 1% level, ** statistically significant at the 5% level, * statistically significant at the 10% level. (6) Absolute values in parentheses below random effects parameters denote standard errors. (7) ICC denotes interclass correlation coefficient

Bold indicate significance at the 1% and 5% for the specific variables

Table 10 Multilevel model results: fully expanded version (Eq. 7/specification 9)

Independent variables	Dependent variables									
	TCR [1]	CRTIER1 [2]	LR [3]	NPLS [4]	PROVLL LOANS [5]	NPLS RESERLL[6]	OPEXPENS OPINCOME [7]	ROA [8]	ROE [9]	LTD [10]
Random effects parameters										
Newgreen × region level	0.000 (0.000)	0.000 (0.000)	0.000 (0.000)	0.000 (0.000)	0.000 (0.000)	0.000 (0.000)	0.000 (0.000)	0.038 (0.048)	0.000 (0.000)	0.000 (0.000)
Newgreen × country level	0.000 (0.000)	0.000 (0.000)	0.000 (0.000)	0.000 (0.000)	0.000 (0.000)	0.000 (0.000)	0.000 (0.000)	0.000 (0.000)	0.000 (0.000)	0.000 (0.000)
Residuals	4.619 (0.218)	4.263 (0.198)	13.732 (0.651)	9.360 (0.442)	0.649 (0.030)	1230.48 (59.031)	62,977.53 (3044.14)	0.456 (0.022)	583.39 (27.335)	338.085 (15.806)
ICC/variance decomposition										
Newgreen × region level %	0.00	0.00	0.00	0.00	0.00	0.00	0.00	0.71	0.00	0.00
Newgreen × country level %	0.00	0.00	0.00	0.00	0.00	0.00	0.00	0.00	0.00	0.00
Individual level (green bank)%	100.00	100.00	100.00	100.00	100.00	100.00	100.00	99.29	100.00	100.00
Diagnostics										
Log likelihood	– 1967.81	– 1987.38	– 2425.92	– 2275.82	– 1101.65	– 4324.59	– 5944.24	– 899.83	– 4193.67	– 3962.49
Wald chi2(1) test: *p*-value	0.0000	0.0000	0.0000	0.0000	0.0000	0.0000	0.0000	0.0000	0.0000	0.0000
Observations	901	927	889	897	916	869	856	874	911	915
No. of groups										
Newgreen × regions	2	2	2	2	2	2	2	2	2	2

(continued)

Table 10 (continued)

Independent variables	Dependent variables									
	TCR [1]	CRTIER1 [2]	LR [3]	NPLS [4]	PROVLL LOANS [5]	NPLS RESERLL[6]	OPEXPENS OPINCOME [7]	ROA [8]	ROE [9]	LTD [10]
Newgreen × countries	3	3	3	3	3	3	3	3	3	3

Notes (1) Method of estimation: MLE. (2) Nesting levels at: region level × green bank, country level × green bank. (3) Constant estimate represents the intercept estimate and can be seen as the grand mean corresponding to each dependent variable [1]–[10]. (4) All independent variables are defined in Tables 1 and 2. (5) Absolute values in parentheses denote *z-statistics* (two-tailed test): *** statistically significant at the 1% level, ** statistically significant at the 5% level, * statistically significant at the 10% level. (6) Absolute values in parentheses below random effects parameters denote standard errors. (7) ICC denotes interclass correlation coefficient

4.3 T-*Tests Results and Discussion*

T-test results are given in Table 6. The hypotheses being tested are presented in Sect. 3, Subsect. 3.3. Specifically, we form six baseline hypotheses (cases); each one of them is decomposed into thirteen null hypotheses (i.e., the number of the variables under investigation). In each case, we keep H_0 if $t < t^*$ and $p > P^*$, where $t^* = 1.96$ and stands for *t*-critical value, and P^* at $a = 0.05$. Otherwise, we reject H_0 and accept Ha.

T-test results show that, in most of the cases examined, there are statistically significant differences in the mean values of CAMEL variables between green and non-green banks during the whole data period as well as in both the pre- and the post-financial crisis period. However, these differences could be driven by other bank characteristics or factors. In addition, the above-mentioned results do not necessarily imply that green banks outperform non-green banks or vice versa. This justifies further analysis to investigate if green ("treatment" group) and non-green banks ("control" group) exhibit indeed differential behavior.

4.4 *Correlation Analysis*

We performed two separate correlation analyses: (a) one with all CAMEL variables and bank size and (b) one with the remaining variables after examining the size of the correlation coefficients. Specifically, in order to avoid possible multicollinearity problems, we decided to exclude all variables[18] with correlation coefficient $r \geq 0.65$. The correlation coefficients of the remaining variables are presented in Table 7,[19] and they are all below 0.50[20] which can be considered as a threshold between comparatively low and moderate and upper correlations (Gujarati, 2004, p. 359).

[18] According to Gujarati (2004, p. 365) [...] "*to drop one of the collinear variables is a rule-of-thumb procedure used to overcome the problem of multicollinearity, albeit this can lead to specification bias*". However, the remaining eleven variables (excluding dummies) and the total number of observations (2805) are sufficiently high, given also the kind of our data (see, e.g., Gujarati, 2004, p. 364). In addition, all three variables (RESERLLLOANS, OPEXPENSTA, and NONINTEXPENSTA) that are excluded from the estimation procedure provide similar information as the remaining variables, since they belong to the same CAMEL's segments, that is, asset quality and management quality.

[19] Due to space limitations, here are reported only the correlation analysis results of the remaining variables.

[20] Excluding: (a) TCR and PROVLLLOANS ($r = 0.556$) and (b) TCR and CRTIER1 ($r = 0.945$) pairwise correlations; in the last case the two variables are used interchangeably in regression analysis.

4.5 Regressions Results and Discussion

4.5.1 Random Effects Model

In Table 8, we present the estimation results of our RE panel data regression models. Specifically, each table shows the results of one CAMEL factor as a dependent variable against the rest of the CAMEL factors, bank size, and dummy variables (global bank, country, financial crisis, green bank, and interaction crisis × green bank) as independent variables. All right-hand side variables (excluding dummy variables) are lagged by one period.[21] Note that the number of observations varies across the regressions, depending on the missing values for the bank-specific variables. In both models' specifications, group variable is bank.

In all cases, i.e., in all combinations of each CAMEL variable as dependent and the rest as independents, the specifications [5]–[9] show the coefficients' estimates when we include successively the dummy variables in the regressions. We add them one-by-one to control for the effect on the dependent variable of global bank, country, crisis, green bank, and interaction *crisis × green.* Specification 9 represents the fully expanded version of our model (Eq. 3) as we include all variables simultaneously (see Table 8).

Next follows the discussion of the estimation results concerning the RE model only.[22]

Our first main hypothesis ("*crisis does not affect CAMEL factors*") is rejected, as we find that the financial crisis has a statistically significant effect, either positive or negative, on all CAMEL factors excluding leverage ratio; this further implies that this capital ratio is a less sensitive one (compared to the risk-based capital ratios) to systemic shocks and financial crises, and hence a more objective measure of banks' capital adequacy, and perhaps a better predictor of banks' solvency. This is in line with Behn et al. (2016), Smith et al. (2017), and Drakos and Malandrakis (2021).

The second main hypothesis ("*bank type does not affect CAMEL factors*"), with the exception of Tier 1 Capital ratio, is accepted for all CAMEL variables; this means that green banks do not differ from their non-green counterparts during the whole data period with respect to all CAMEL variables, excluding Tier 1 Capital ratio (which is lower than the corresponding ratio of non-green banks but only in the pre-crisis period, since the effect of the crisis on green banks' Tier 1 Capital ratio is positive as explained below).

Finally, our third main hypothesis ("*crisis has no impact on green banks*") is rejected for the Total Capital and Tier 1 Capital ratios, but it is confirmed for the rest of the CAMEL variables; this finding indicates that (a) green banks' risk-adjusted capital ratios were improved during and after the global financial crisis period (green

[21] In all models, explanatory variables are lagged by one period to avoid possible endogeneity issues.

[22] Considering the results obtained for the specification 9 of our RE model.

banks performed better relative to non-green banks with respect to this ratio), and (b) green banks do not differ from their non-green counterparts in the post-crisis period with respect to all other CAMEL ratios (i.e., the leverage ratio, and the asset quality, management ability, earning ability, and liquidity factors).

As for the control dummy variables *global* and *bank size*, our results provide evidence that (a) the global presence of a bank has a negative effect on provisions (global banks, whether green or not, exhibit a lower provisions ratio than non-global banks) and on Loan-To-Deposit[23] ratios only, in both the pre- and the post-crisis period; (b) bank size generally is irrelevant, whether a bank is green or not, during the whole data period.[24] Moreover, decomposing the effects of the above-mentioned control variables separately for the pre- and the post-financial crisis period, we find that (a) the global position of bank has a significant positive impact on Tier 1 Capital ratio in the post-crisis period and on ROA ratio in the pre-crisis period, and (b) bank size has affected positively NPLs Reserves ratio in the pre-crisis period and negatively operational expenses and ROA ratios in the post-crisis period.

Considering the interaction of the CAMEL bank-specific variables with each other, regarding green and non-green banks in both the pre- and post-crisis period,[25] our results exhibit that:

(a) Regarding capital adequacy: the lower the liquidity, the lower the capital adequacy; and the higher the equity return, the higher the Tier 1 Capital[26] and leverage ratios; since ROE ratio is the only CAMEL factor that affects materially leverage ratio, an additional indication exists that this non-risk-adjusted capital ratio remains unaffected not only by financial crisis but also by the other CAMEL factors, enhancing our prior finding that this ratio constitutes a more objective measure of banks' capital adequacy.
(b) Considering asset quality indicators: (i) higher operational expenses and equity return ratios imply higher[27] and lower NPLs ratios, respectively; (ii) a better Total Capital adequacy ratio implies a lower Provisions' ratio (a result which might require further investigation) while higher operational costs and higher profitability imply higher provisions (the latter result also needs further research); (iii) Provisions' ratio has a negative influence on NPLs Reserves ratio, bank profitability has a mixed effect on NPLs Reserves ratio (e.g., ROA increases and ROE decreases this ratio); (iv) lower liquidity implies a higher NPLs Reserves ratio.

---

[23] Although this effect comes after countries inclusion, it is statistically significant mainly at the 10% level.

[24] Excluding green banks in the after-crisis period, where bank size was found to have a negative impact on their liquidity, although this is a rather weak relationship given the relatively low levels of statistical significance.

[25] Note that we limit our discussion to mentioning only the cases where the corresponding estimate is statistically significant in all model's specifications, as well as in special cases.

[26] Note that the possible impact of the Tier 1 Capital ratio to the rest of the CAMEL variables is not examined, since we have excluded from the estimation procedure this CAMEL ratio as independent variable, because of the high degree of correlation with Total Capital ratio.

[27] Albeit not significant in size, considering the magnitude of the relevant estimate.

(c) Regarding management ability factors: better capitalization in terms of Total Capital ratio implies significant lower operational costs and hence a better management ability, and lower bank liquidity implies higher operational expenses ratio and hence inferior management ability.
(d) As for the profit ability ratios: a higher Provisions' ratio implies higher profitability in terms of the ROA ratio only (a result that might require further research), lower management ability implies lower bank profitability[28] and, finally, a higher NPLs Reserves ratio exerts a negative impact on banks profitability, but only in the pre-crisis period.
(e) Regarding liquidity: Loan-To-Deposit increases when ROE and leverage ratio rise (although in this last case the effect becomes insignificant after countries introduction and turns again into significant but at the 10% level in the after-crisis period) a somewhat controversial result, as an opposite effect was expected.

To sum up, our results suggest that, at present, green banks may not be very different from non-green banks. Our empirical estimations show that green banks performed better in terms of the two risk-based capital ratios (Total Capital and Tier 1 Capital) during the global financial crisis, though this result does not necessarily mean that they are generally better capitalized than their non-green counterparts. The better performance of green banks with respect to these two risk-adjusted capital adequacy ratios, during and after the crisis, might be the result of higher capital injections by the state and/or other factors such as variations across countries and across green banks of different sizes. On the other hand, green banks they do not exhibit statistically significant differences from non-green banks (for instance, they do not surpass non-green banks in terms of leverage ratio, asset quality, management quality, earning ability, and liquidity, both in the pre- and the post-crisis period). Finally, considering the impact of CAMEL variables between them, examining both bank types during the whole data period (1999–2015), we derive various interactions, with the most notable findings being that (a) leverage ratio is probably a more objective measure of both green and non-green banks' capital adequacy, since it remains unaffected by the vast majority of CAMEL factors, and this holds not only in the pre- but also in the post-crisis period, (b) better liquidity implies better capital adequacy, and (c) NPLs ratio does not affect the vast majority of the CAMEL variables (excluding ROE and only after the crisis), a result that may be attributed to various reasons, for example the effect of NPLs in one or more region/s (e.g., Europe) being absorbed by the effect of NPLs in another region/s (e.g., North America, Oceania), and/or to the different regulatory treatment of NPEs and NPLs across different countries and jurisdictions.

[28] However, in the ROA ratio case the magnitude of the relevant estimate—despite the high level of statistical significance—is very small, while in the ROE ratio case the level of significance drops to the 10% level after the introduction of the crisis dummy.

4.5.2 Multilevel Model

In Tables 9 and 10, we present the estimation results of our multilevel regression model. Note that the first part of Table 9 follows the procedure presented in Subsect. 4.5.1. Again, all right-hand side variables (excluding dummy variables) are lagged by one period.[29]

The application of the multilevel model produces twofold results. First, regarding our hypotheses, we can infer that (a) the crisis still constitutes an important factor affecting almost all CAMEL factors excluding leverage ratio as before, plus Operational Expenses/Operating Income and Loan-To-Deposit ratios (i.e., these last two CAMEL factors remained relatively unaffected, as some countries belonging in certain regions were hit less by the 2007–2008 financial crisis); (b) green banks do not differ (e.g., they do not perform better than conventional banks) in terms of all CAMEL variables (note previous result concerning Tier 1 Capital ratio); (c) the crisis has no impact on green banks regarding all CAMEL ratios (excluding Tier 1 Capital ratio which is marginally significant at the 10% level). By examining the random effects parameters of our first multilevel model (see Eq. 6 and Table 9), and specifically the corresponding standard errors and the calculated interclass correlation coefficients (ICC), it seems that neither region nor country exerts a significant influence on CAMEL variables' values (it is rather a matter of bank characteristics, e.g., of the financial risk profile at the individual level) with the higher percentages being that for TCR at the country level (1.27%) and for ROA at the region level (2.12%). Moreover, we can infer that bank level accounts for more than 90% considering all CAMEL variables variance. By examining the ICC values and the variance decomposition estimates of green bank with country and of green bank with region (see Eq. 7 and Table 10), all are accounting for less than 0.5–1%, i.e., the country and region effect with respect to green banks is negligible.

So we conclude that, with the data set used (165 banks from 38 countries from six world regions) and for the specific time period (1999–2015), the application of a multilevel (or hierarchical) model shows that the higher proportion of all CAMEL variables variance, irrespective of bank type, is due to the level of the bank, suggesting that intrinsic bank characteristics are probably responsible for a high significant portion of each variable variance per bank per year, and that country and region effects are rather unimportant.

5 Conclusions and Policy Implications

Our main results exhibit that green banks, whether global or not, generally do not differ from (e.g., do not outperform) their non-green counterparts in terms of CAMEL ratios before and after the financial crisis. We also find that crisis has equally affected

[29] In all models, explanatory variables are lagged by one period to avoid possible endogeneity issues.

the ratios of both bank-types, either positively or negatively: (a) a positive effect on capital adequacy (excluding leverage ratio thus indicating that this capital ratio is less sensitive to systemic shocks and financial crises), asset quality (excluding NPLs ratio) and management quality, and (b) a negative effect on earnings ability and on liquidity. When we interact the financial crisis with green banks, our results show that green banks exhibit a statistically significant higher Total Capital and Tier 1 Capital ratio solely during and after the financial crisis. With reference to the rest of the CAMEL factors, our results demonstrate that there are no statistically significant differences between the two groups in the post-crisis period, irrespective of global presence or not. The empirical application of a multilevel model confirms most of the above-mentioned results, although we find that apart from the leverage ratio, the Operational Expenses/Operating Income (management quality proxy), and the Loan-To-Deposit (liquidity proxy) also remain unaffected by the financial crisis, a result that needs further investigation. Variance decomposition results show that neither country nor region has any significant effect on CAMEL variables values (it is rather a matter of bank characteristics, either green or non-green).

In terms of policy implications, our main findings provide evidence that green banks are not necessarily better in terms of credit risk, so that lower capital requirements and a different regulatory regime for green banks could not be justified at the moment as it could lead to financial instability and perhaps to a new banking crisis. This is in line with Boot and Schoenmaker (2018) who argue that banks will need not less but more capital in the new era. Moreover, green banks presumably are not green enough, given that the exposures of some leading banking sectors of the world as well as global banks in environmentally unfriendly sectors are still high and increasing (from 1.4 trillion USD in December 2014 to 1.9 trillion USD in December 2018) (see, e.g., Banking on Climate Change, 2019; Nieto, 2017). In addition, favorable measures for green loans, such as a "green-supporting factor", and a "brown-penalizing factor", must not be adopted yet without sufficient evidence on green loans risk level and their contribution to environment protection, and green investments' returns, as they may result in the reduction of banks' capital and the creation of a general and/or systematic financial instability (see, e.g., Berenguer et al., 2020; Dafermos et al., 2018; Thomä & Gibhart, 2019). A sudden implementation of a green regulatory framework may result into regulatory arbitrage. An interim measure such as the "Sustainable Finance Supporting Factor", proposed by the European Banking Federation (2021), and a gradual movement from polluting to eco-friendly projects and sectors is advised, in order to avoid significant direct and indirect negative effects such as a sudden increase in energy costs or an abrupt depreciation of fossil fuel reserves that could contribute to a generalized financial instability (see, e.g., Batten et al., 2016; Manninen & Tiililä, 2020; Nieto, 2017) as a result, inter alia, of "stranded assets" (Papandreou, 2019; Xepapadeas, 2021). Finally, on the basis of our results and taking into account the related literature, we formulate certain propositions: first, the establishment of a globally accepted set of classification criteria for the categorization of a bank as green or not; second, regulators must (a) set out new capital requirements for credit risk, operational risk, and market risk by taking into account the new risks imposed by climate change (i.e., transition, physical, and liability risks); (b) establish a new regulatory framework for NPEs and NPLs that will take into account issues such as NPLs and NPEs distribution on high and low polluting sectors, the potential impact of physical risk and transition risk on green and non-green banks' loan portfolios; (c) consider the development of environmentally adjusted capital adequacy ratios, putting special emphasis on the leverage ratio.

The main limitation of our analysis was the unavailability of data related to green loans portfolio and the proportion of green investments per bank per year. Finally, there is a range of issues that could be explored in a future research, such as the enrichment of the data set with more banks from additional countries, a more up-to-date time period (i.e., from 2016 onwards), the incorporation in the analysis of the GSF and BPF (given data availability), and of some macroeconomic and environmental variables (e.g., at country level).

Acknowledgements We would like to thank the two anonymous reviewers for their constructive comments that contributed to the improvement of this paper. We also wish to thank the participants of the 2021 Annual Conference of the Scottish Economic Society and of the 10th International Conference of the Financial Engineering and Banking Society (FEBS) 2021 for their helpful comments and suggestions. Finally, we want to thank Professor C. Zopounidis and Mrs. A. Liadakis for their kind support.

References

Afroj, F. (2022). Financial strength of banking sector in Bangladesh: A CAMEL framework analysis. *Asian Journal of Economics and Banking*. https://doi.org/10.1108/AJEB-12-2021-0135

Anastasiou, D., Louri, H., & Tsionas, M. (2019). Nonperforming loans in the euro area: Are core–periphery banking markets fragmented? *International Journal of Finance & Econonics, 24*, 97–112. https://doi.org/10.1002/ijfe.1651

Baltagi, H. D. (2005). *Econometric analysis of panel data* (3rd ed.). Wiley.

Bank for International Settlements (BIS). (2009, June 29). *79th annual report*, Basel.

Banking on Climate Change. (2019, March 20). *Fossil fuel finance report card 2019*. https://www.ran.org/bankingonclimatechange2019. Accessed April 29, 2019.

Basel Committee on Banking Supervision (BCBS). (2013, July). *Global systemically important banks: Assessment methodology and the higher loss absorbency requirement*. Bank for International Settlements.

Basel Committee on Banking Supervision (BCBS). (2017, March). *Global systemically important bank—Revised assessment framework*. Bank for International Settlements.

Basel Committee on Banking Supervision (BCBS). (2019). *Leverage ratio requirements for global systemically important banks*. Bank for International Settlements. https://www.bis.org/basel_framework/chapter/LEV/40.htm?inforce=20220101. Accessed August 24, 2019.

Basten, M., & Serrano, A. S. (2019). European banks after the global financial crisis: A new landscape. *Journal of Banking Regulation, 20*, 51–73. https://doi.org/10.1057/s41261-018-0066-3

Batten, S., Sowerbutts, R., & Tanaka, M. (2016, May). *Let's talk about the weather: The impact of climate change on central banks* [Staff Working Paper No. 603]. Bank of England.

Baur, G. D. (2012). Financial contagion and the real economy. *Journal of Banking & Finance, 36*, 2680–2692. https://doi.org/10.1016/j.jbankfin.2011.05.019

Beck, T., Demirgüc-Kunt, A., & Merrouche, O. (2013). Islamic vs. conventional banking: Business model, efficiency and stability. *Journal of Banking & Finance, 37*, 433–447. https://doi.org/10.1016/j.jbankfin.2012.09.016

Behn, M., Haselmann, R., & Wachtel, P. (2016). Procyclical capital regulation and lending. *Journal of Finance, LXXI*(2), 919–955. https://doi.org/10.1111/jofi.12368

Benedikter, R. (2011). *Social banking and social finance answers to the economic crisis.* Springer. https://doi.org/10.1007/978-1-4419-7774-8

Berenguer, M., Cardona, M., & Evain, J. (2020, March). *Integrating climate-related risks into banks' capital requirements*, Paris. Institute For Climate Economics (I4CE).

Berger, N. A., & DeYoung, R. (1997). Problem loans and cost efficiency in commercial banks. *Journal of Banking & Finance, 21*, 849–870. https://doi.org/10.1016/S0378-4266(97)00003-4

Berger, N. A., & Roman, A. R. (2020). Methodologies used in most of the TARP empirical studies (Chap. 5). In *TARP and other bank bailouts and bail-ins around the world: Connecting wall street, main street, and the financial system* (pp. 177–185). Academic Press. https://doi.org/10.1016/B978-0-12-813864-9.00005-7

Boot, A., & Schoenmaker, D. (2018, January 17). *Climate change adds to risk for banks, but EU lending proposals will do more harm than good.* University of Amsterdam, UvA-DARE (Digital Academic Repository), Council on Economic Policies (CEP). https://www.cepweb.org/climate-change-adds-to-risk-for-banks-but-eu-lending-proposals-will-do-more-harm-than-good/. Accessed May 06, 2018. https://pure.uva.nl/ws/files/31245768/Climate_change_adds_to_risk_for_banks_but_EU_lending_proposals_will_do_more_ha.pdf. Accessed May 28, 2019.

Boston Common Asset Management. (2015). *Are banks prepared for climate change?* [Impact Report 2015]. http://news.bostoncommonasset.com. Accessed March 13, 2018.

Brambor, T., Clark, W. R., & Golder, M. (2006). Understanding interaction models: Improving empirical analyses. *Political Analysis, 14*(1), 63–82. http://www.jstor.org/stable/25791835

Braumoeller, B. F. (2004). Hypothesis testing and multiplicative interaction terms. *International Organization, 58*(4), 807–820. https://www.jstor.org/stable/3877804

Ceres. (2021). *About us.* https://www.ceres.org/about-us. Accessed October 27, 2021.

Cetorelli, N., & Goldberg, S. L. (2010, May). *Global banks and international shock transmission: Evidence from the crisis* [Federal Reserve Bank of New York Staff Reports No. 446].

Chernykh, L., & Kotomin, V. (2022). Risk-based deposit insurance, deposit rates and bank failures: Evidence from Russia. *Journal of Banking & Finance, 138*, 1–13. https://doi.org/10.1016/j.jbankfin.2022.106483

Chiaramonte, L., Croci, E., & Poli, F. (2015). Should we trust the Z-score? Evidence from the European Banking Industry. *Global Finance Journal, 28*, 111–131. https://doi.org/10.1016/j.gfj.2015.02.002

Choi, I. (2001). Unit root tests for panel data. *Journal of International Money and Finance, 20*, 249–272. https://doi.org/10.1016/S0261-5606(00)00048-6

Christopoulos, G. A., Mylonakis, J., & Diktapanidis, P. (2011). Could Lehman brothers' collapse be anticipated? An examination using CAMELS rating system. *International Business Research, 4*(2), 11–19. https://doi.org/10.5539/ibr.v4n2p11

Clark, T., & Linzer, D. (2015). Should I use fixed or random effects? *Political Science Research and Methods, 3*(2), 399–408. https://doi.org/10.1017/psrm.2014.32

Coalition for Green Capital (CGC). (2017). *Growing clean energy markets with green bank financing* [White Paper]. http://coalitionforgreencapital.com. Accessed July 31, 2017.

Coalition for Green Capital (CGC). (2019). *What is a green bank?* http://coalitionforgreencapital.com/whats-a-green-bank-html/. Accessed December 01, 2019.

Cogan, G. D. (2008, January). *Corporate governance and climate change: The banking sector* [Ceres Report]. https://era.library.ualberta.ca/items/ab2b6b4d-4414-460b-ad7a-cf39c40954c7/view/d0af3bde-bc35-4ce3-a0d3-73d2bfec5ab5/ceres_climate_change_banking_report2008.pdf. Accessed October 23, 2018.

Cole, A. R., & White, J. L. (2017). When time is not on our side: The costs of regulatory forbearance in the closure of insolvent banks. *Journal of Banking & Finance, 80*, 235–249. https://doi.org/10.1016/j.jbankfin.2017.03.010

Dafermos, Y., & Nikolaidi, M. (2021). How can green differentiated capital requirements affect climate risks? A dynamic macrofinancial analysis. *Journal of Financial Stability, 54*, 100871, 1–27. https://doi.org/10.1016/j.jfs.2021.100871

Dafermos, Y., Nikolaidi, M., & Galanis, G. (2018, July). *Can green quantitative easing (QE) reduce global warming?* [Policy Brief]. Greenwich Political Economy Research Centre (GPERC).

Danmarks Nationalbank. (2019, December 2). Climate change can have a spillover effect on financial stability. *Analysis*, No. 26.

De Bondt, G., & Vermeulen, P. (2021). Business cycle duration dependence and foreign recessions. *Scottish Journal of Political Economy, 68*(1), 1–19. https://doi.org/10.1111/sjpe.12261

De Haas, R., & Van Lelyveld, I. (2014). Multinational banks and the global financial crisis: Weathering the perfect storm? *Journal of Money, Credit and Banking, 46*(1), 333–364. https://doi.org/10.1111/jmcb.12094

Doumpos, M., & Zopounidis, C. (2010). A multicriteria decision support system for bank rating. *Decision Support Systems, 50*, 55–63. https://doi.org/10.1016/j.dss.2010.07.002

Doumpos, M., Hasan, I., & Pasiouras, F. (2017). Bank overall financial strength: Islamic versus conventional banks. *Economic Modelling, 64*, 513–523. https://doi.org/10.1016/j.econmod.2017.03.026

Drakos, K., & Malandrakis, I. (2021). Global versus non-global banks: A capital ratios-based analysis. *Journal of Central Banking Theory and Practice, 10*(2), 5–22. https://doi.org/10.2478/jcbtp-2021-0011

Dunz, N., Naqvi, A., & Monasterolo, I. (2021). Climate sentiments, transition risk, and financial stability in a stock-flow consistent model. *Journal of Financial Stability, 54*, 2–37. https://doi.org/10.1016/j.jfs.2021.100872

Eckstein, D., Künzel, V., Schäfer, L., & Winges, M. (2020). *Global climate risk index 2020* [Briefing Paper], December 2019. Germanwatch e.V.

Equator Principles (EP). (2021a). *The equator principles*. http://equator-principles.com/about/. Accessed October 27, 2021.

Equator Principles (EP). (2021b). *EP association members & reporting*. http://equator-principles.com/members-reporting/. Accessed October 27, 2021.

European Banking Authority (EBA). (2019). *Sustainable finance*. https://eba.europa.eu/financial-innovation-and-fintech/sustainable-finance. Accessed April 17, 2019.

European Banking Federation (EBF). (2021). *Supporting factor*. https://www.ebf.eu/sustainable-finance/supporting-factor/

European Commission. (2018). *Action plan: Financing sustainable growth*. COM (2018) 97 final, 8.3.2018, Brussels. https://eur-lex.europa.eu/legal-content/EN/TXT/PDF/?uri=CELEX:52018DC0097&from=EN. Accessed April 17, 2019.

Federal Deposit Insurance Corporation (FDIC). (2018). *Basic examination concepts and guidelines*, 03/18. https://www.fdic.gov/regulations/safety/manual/section1-1.pdf. Accessed July 01, 2018.

Fillat, L. J., Garetto, S., & Götz, M. (2013, December 7). *Multinational banks*. https://cepr.org/sites/default/files/Fillat.pdf. Accessed March 23, 2018.

Financial Stability Board (FSB). (2016). *2016 list of global systemically important banks (G-SIBs)*. https://www.fsb.org/2016/11/2016-list-of-global-systemically-important-banks-g-sibs/

G-20 Green Finance Study Group. (2016, September 5). *G20 green finance synthesis report*. http://unepinquiry.org/wp. Accessed March 30, 2018.

Gelman, A., & Hill, J. (2007). *Data analysis using regression and multilevel/hierarchical models*. Cambridge University Press.

Gilbert, A. R., Meyer, P. A., & Vaughan, D. M. (2000). *The role of a CAMEL downgrade model in bank surveillance* [Working Paper 2000–021A]. Federal Reserve Bank of St. Louis.

Gilleo, A., Stickles, B., & Kramer, C. (2016, September). *Green bank accounting: Examining the current landscape and tallying progress on energy efficiency* [Report F1602]. American

Council for an Energy-Efficient Economy. http://aceee.org/research-report/f1602. Accessed July 31, 2017.

Global Alliance for Banking on Values (GABV). (2021a). *About*. http://www.gabv.org/about-us. Accessed October 27, 2021.

Global Alliance for Banking on Values (GABV). (2021b). *List of the GABV member banks*. https://www.gabv.org/wp-content/uploads/List-of-GABV-Members-August-2021.pdf. Accessed October 27, 2021.

Gopalan, K. Y. (2010). Earliest indicator of bank failure is deterioration in earnings. *Central Banker*, Spring 2010. Federal Reserve Bank of St. Louis.

Gujarati, D. (2004). *Basic econometrics* (4th ed.). McGraw-Hill Companies.

Hale, G., Kapan, T., & Minoiu, C. (2016). *Crisis transmission through the global banking network* [IMF Working Paper WP/16/91].

Hirtle, B., Konver, A., & Plosser, M. (2020). The impact of supervision on bank performance. *The Journal of Finance, LXXV*(6), 2765–2808. https://doi.org/10.1111/jofi.12964

Imbens, G., & Wooldridge, J. (2007). Difference-in-differences estimation. In *Lecture Notes 10-Summer '07*. NBER. https://www.nber.org/lecture/summer-institute-2007-methods-lecture-difference-differences-estimation

Intergovernmental Panel on Climate Change (IPCC). (2021). Summary for policymakers: B. Possible climate futures. In V. Masson-Delmotte, P. Zhai, A. Pirani, S. L. Connors, C. Péan, S. Berger, N. Caud, Y. Chen, L. Goldfarb, M. I. Gomis, M. Huang, K. Leitzell, E. Lonnoy, J. B. R. Matthews, T. K. Maycock, T. Waterfield, O. Yelekçi, R. Yu, & B. Zhou (Eds.), *Climate Change 2021: The Physical Science Basis Contribution of Working Group I to the Sixth Assessment Report of the Intergovernmental Panel on Climate Change, IPCC AR6 WGI*. Cambridge University Press. In Press.

International Finance Corporation (IFC)—World Bank. (2010). *Telling our story: Climate change: Filling the financing gap*, 59447. World Bank Group. http://documents.worldbank.org/curated/en/440331468337913307/Telling-our-story-climate-change-filling-the-financing-gap. Accessed March 28, 2016, May 06, 2018.

International Finance Corporation (IFC)—World Bank & Sustainable Banking Network (SBN). (2017). *Greening the banking system—Experiences from the sustainable banking network (SBN) (input paper for the G20 green finance study group)*. https://www.ifc.org/wps/wcm/connect/413da3f3-f306-4e99-8eee-679768463130/SBN_PAPER_G20_02102017.pdf?MOD=AJPERES. Accessed May 06, 2018.

Jeucken, M. (2001). *Sustainable finance and banking: The financial sector and the future of the planet*. Earthscan Publications Ltd.

Kalemli-Ozcan, S., Papaioannou, E., & Perri, F. (2013). Global banks and crisis transmission. *Journal of International Economics, 89*, 495–510. https://doi.org/10.1016/j.jinteco.2012.07.001

Kayo, K. E., & Kimura, H. (2011). Hierarchical determinants of capital structure. *Journal of Banking & Finance, 35*, 358–371. https://doi.org/10.1016/j.jbankfin.2010.08.015

Klomp, J. (2014). Financial fragility and natural disasters: An empirical analysis. *Journal of Financial Stability, 13*, 180–192. https://doi.org/10.1016/j.jfs.2014.06.001

Kupiec, P., Lee, Y., & Rosenfeld, C. (2017). Does bank supervision impact bank loan growth? *Journal of Financial Stability, 28*, 29–48. https://doi.org/10.1016/j.jfs.2016.11.006

Laeven, L., & Valencia, F. (2012, June). *Systemic banking crises database: An update* [IMF Working Paper WP/12/163]. International Monetary Fund (IMF).

Lamperti, F., Bosetti, V., Roventini, A., & Tavoni, M. (2019). The public costs of climate-induced financial instability. *Natural Climate Change, 9*, 829–833. https://doi.org/10.1038/s41558-019-0607-5

Makridou, G., Doumpos, M., & Galariotis, E. (2019). The financial performance of firms participating in the EU emissions trading scheme. *Energy Policy, 129*, 250–259. https://doi.org/10.1016/j.enpol.2019.02.026

Manninen, O., & Tiililä, N. (2020). *Could the green supporting factor help mitigate climate change?* Bulletin, Bank of Finland.

Moosa, I. (2010). The myth of too big to fail. *Journal of Banking Regulation, 11*, 319–333. https://doi.org/10.1057/jbr.2010.15

Niepmann, F. (2011, December). *Banking across borders* [DNB Working Paper No. 325]. De Nederlandsche Bank (DNB).

Nieto, J. M. (2017, October). *Banks and environmental sustainability: Some financial stability reflections* [Working Paper]. International Research Centre on Cooperative Finance (IRCCF). http://www.financecoop.hec.ca. Accessed October 30, 2018.

NGFS. (2018). *Charter of the central banks and supervisors network for greening financial system (NGFS)*. https://www.ngfs.net/en. Accessed March 28, 2019.

NGFS. (2019, 2021). *About us*. https://www.ngfs.net/en. Accessed March 28, 2019, October 27, 2021.

Nguyen, H. A., Nguyen, T. H., & Huong, T. P. (2020). Applying the CAMEL model to assess performance of commercial banks: Empirical evidence from Vietnam. *Banks and Bank Systems, 15*(2), 177–186. https://doi.org/10.21511/bbs.15(2).2020.16

OECD. (2015, December). Green investments banks. *OECD Policy Perspectives*. http://www.oecd.org/environment/green-investment-banks.htm. Accessed March 28, 2016.

Papandreou, A. (2019, November). *Stranded assets and the financial system* [Working Paper No. 272]. Bank of Greece.

Papanikolaou, N. (2018). To be bailed out or to be left to fail? A dynamic competing risks hazard analysis. *Journal of Financial Stability, 34*, 61–85. https://doi.org/10.1016/j.jfs.2017.11.005

Peria Martinez, S. M., & Schmukler, L. S. (2001). Do depositors punish banks for bad behavior? Market discipline, deposit insurance, and banking crises. *The Journal of Finance, LVI*(3), 1029–1051. https://doi.org/10.1111/0022-1082.00354

Sartzetakis, S. E. (2019, March). *Green bonds as an instrument to finance low carbon transition* [Working Paper No. 258]. Bank of Greece.

Sharfman, P. M., & Fernando, S. C. (2008). Environmental risk management and the cost of capital. *Strategic Management Journal, 29*, 569–592. https://doi.org/10.1002/smj.678

Smith, A. J., Grill, M., & Lang, J. H. (2017). *The leverage ratio, risk-taking and bank stability* [Working Paper Series No. 2079]. European Central Bank.

Swindle, S. C. (1995). Using CAMEL ratings to evaluate regulator effectiveness at commercial banks. *Journal of Financial Services Research, 9*, 123–141. https://doi.org/10.1007/BF01068074

Thomä, J., & Gibhart, K. (2019). Quantifying the potential impact of a green supporting factor or brown penalty on European banks and lending. *Journal of Financial Regulation and Compliance, 27*(3), 380–394. https://doi.org/10.1108/JFRC-03-2018-0038

UNEP Inquiry. (2016, September). *Definitions and concepts: Background note* [Inquiry Working Paper 16/13]. http://unepinquiry.org/wp-content/uploads/2016/09/1_Definitions_and_Concepts.pdf. Accessed April 17, 2018.

United Nations Environment Programme—Finance Initiative (UNEP FI). (2021a). *About us*. https://www.unepfi.org/about/

United Nations Environment Programme—Finance Initiative (UNEP FI). (2021b). *Our members*. http://www.unepfi.org/members/

University of Cambridge Institute for Sustainability Leadership (CISL). (2018). *Banking environment initiative*. https://www.cisl.cam.ac.uk/business-action/sustainable-finance/banking-environment-initiative. Accessed October 06, 2018, October 27, 2021.

U.S. Alliance for Sustainable Finance. (2019). *U.S. alliance for sustainable finance*. https://go.bloomberg.com/events/usasf/. Accessed April 26, 2019.

Van den End, W. J. (2013). *A macroprudential approach to address liquidity risk with the loan-to-deposit ratio* [DNB Working Paper No. 372].

Whalen, G. (2005, May). *A hazard model of CAMELS downgrades of low-risk community banks* [Economics Working Paper 2005-1]. Office of the Comptroller of the Currency (OCC).
Wooldridge, M. J. (2002). *Econometric analysis of cross section and panel data.* The MIT Press.
Xepapadeas, A. (2021). Climate change and the financial system: A note. *Journal of Industrial and Business Economics, 48*, 5–13. https://doi.org/10.1007/s40812-020-00158-7

Measuring Corporate Gender Diversity and Inclusion with UW-TOPSIS and Linguistic Intervals

Vicente Liern and Blanca Pérez-Gladish

Abstract The objective of this work is to propose fuzzy adequacy indicators to measure the degree of gender diversity in firms. The construction of these indicators will be based on an extension of the Unweighted Technique for Order of Preference by Similarity to Ideal Solution (UW-TOPSIS). This multiple criteria decision-making (MCDM) method simultaneously minimizes the distance to a positive ideal solution and maximizes distance to a negative ideal solution. The positive ideal solution is composed of the best value of each criterion, and the negative ideal solution is composed of the worst values of the decision criteria. The method provides a cardinal ranking of alternatives based on a relative proximity index to the positive ideal solution. In our proposal, the relative importance of the diversity and inclusion decision criteria will be described by means of linguistic labels which will be transformed into intervals on the real line. The main features and advantages of this approach will be illustrated with a real problem where a set of Finnish companies will be assessed based on their degree of adequacy in terms of gender inclusion and diversity.

Keywords MCDM · UW-TOPSIS · Linguistic intervals · Weights · Diversity · Inclusion · Ranking

---

V. Liern
Departamento de Matemáticas para la Economía y la Empresa, Universidad de Valencia, avda Tarongers s/n, 46022 Valencia, Spain
e-mail: vicente.liern@uv.es

B. Pérez-Gladish (✉)
Departamento de Economía Cuantitativa, Universidad de Oviedo, avda del Cristo s/n, 33006 Oviedo, Spain
e-mail: bperez@uniovi.es

C. Zopounidis et al. (eds.), *Operational Research Methods in Business, Finance and Economics*, Lecture Notes in Operations Research,
https://doi.org/10.1007/978-3-031-31241-0_4

1 Introduction

A recently published review of academic research on the impact of diversity and inclusion (D&I) in the workplace has analyzed its relationship with business performance, risk management, and conduct outcomes (Whiting, 2021). The author, after a revision of 169 studies published by academic researchers, consultancies, government, and trade associations, concludes that gender diversity in senior leadership can be associated with positive financial performance, especially when there is a "critical mass", at least 30%, of women on board. However, there is no clear evidence of a causal relation. The evidence is stronger when the relation and causality are analyzed for board gender diversity and risk management. Almost all the 21 studies analyzed by Whiting (2021) find a positive relation between board gender diversity and risk management although only nine of them demonstrate a direct causality. Some of the studies argue that women are more risk averse than men and tend to be found on the boards of less risky firms. However, other authors argue that "this stereotyping does not hold for women who embark on a managerial career, especially in the case of financial services" (Whiting, 2021).

Firms with gender-diverse boards tend to be also more creative, innovative, and able to better solve problems (Torchia et al., 2011; Vafaei et al., 2020) causing a positive impact on their financial performance (Deszö & Ross, 2012; Richard et al., 2003; Whiting, 2021). Gender diversity seems also to have a positive impact on board meeting attendance and financial information transparency and disclosure (Whiting, 2021).

Few studies can be found trying to analyze the impact of gender inclusion on business performance. Although often used interchangeably with diversity, inclusion is a different concept. Diversity in the workplace means that firms employ a diverse team of workers reflecting the society in which the firm exists and operates. Inclusion goes beyond diversity, being defined by the Society for Human Resource Management as "(…) the achievement of a work environment in which all individuals are treated fairly and respectfully, have equal access to opportunities and resources, and can contribute fully to the organization's success" (SHRM, 2022). Due to the nature of this definition, measuring inclusion represents an important challenge. The literature review conducted by Whiting (2021) reveals a lack of consistent measurement data. This author concludes that most of the academic researchers use existing secondary diversity and inclusion data over conducting primary research. The data in most of the analyzed studies are rarely complete because some important variables and characteristics cannot be easily collected, meaning they cannot be included as control variables when analyzing the impact of diversity on performance. The demonstration of causality also requires long times data series which, for diversity and inclusion, are not available.

In addition, the existing diversity and inclusion indicators suffer from some methodological problems common to any overall performance measure. The construction of composite indicators implies several problems concerning collection of data, selection of criteria and individual indicators, normalization of the data, determination of the relative importance (weighting) of criteria and indicators, and aggregation and comparison of overall performance of the alternatives or options. In this work, we will focus on the problematic related to the determination of the criteria weights. Weights can be determined objectively or subjectively, depending on the characteristics of the real decision problem to be solved. Ouenniche et al. (2018) present a review of both types of weighting schemes highlighting the advantages and disadvantages of objective and subjective approaches. In general, the use of subjective weighting schemes is more controversial although is common in the context of TOPSIS-based approaches where decision-makers determine the relative importance of decision criteria based on their own experiences, knowledge, and perception of the problem. Several works including interesting reviews of subjective weighting methods are Barron and Barrett (1996) and Hobbs (1980) and more recently, Alemi-Ardakani et al. (2016), Eshlaghy and Radfar (2006) and Németh et al. (2019). The review of the literature shows that sometimes, the decision-maker cannot give consistent judgments under different weighting schemes and the weighting process itself is essentially context dependent (Watröbski et al., 2019). Therefore, determining reliable subjective weights is a difficult problem and can affect final decisions (Deng et al., 2000). The proposed method in this paper will show how it is possible to obtain similar results to those obtained by a well-known rating agency with a more general weighting scheme without the necessity of the a priori exact numerical establishment of the relative importance of the decision criteria. With this, we will avoid one of the most controversial questions in the construction of global or synthetic indicators.

Equileap is one of the leading EU gender diversity data providers. They research and rank more than 3500 public companies all over the world. Equileap evaluates firms based on 19 diversity and inclusion criteria organized into four main dimensions: gender balance in leadership and workforce, fair remuneration, policies promoting gender equality, and commitment, transparency, and accountability. This organization ranks firms based on a global diversity and inclusion score. Equileap does not provide public information about the relative importance given to the individual indicators and dimensions used to globally score the companies.

Our evaluation framework will rely on a multiple criteria decision analysis approach, Unweighted Technique for Order Preference by Similarity to Ideal Solution, and UW-TOPSIS developed by Liern and Pérez-Gladish (2022). This method allows us to consider the complex multidimensional character of decision problems avoiding some of the difficulties related to the determination of the relative importance of these multiple dimensions. The novel contribution of this paper is related to the type of required information regarding the importance of the decision criteria in the aggregation process leading to the ranking of the firms. In the UW-TOPSIS framework, weights are treated as unknown variables in the optimization problem which determines the worst and best possible relative proximity of each decision alternative to the positive ideal solution (PIS). In this work, we give the decision-maker the opportunity of assessing the importance of the decision criteria using linguistic terms that are transformed into numerical intervals included in the optimization problem.

In what follows we will present the main characteristics of the classical TOPSIS approach followed by a description of the UW-TOPSIS algorithm developed by Liern and Pérez-Gladish (2022). Once the methodological framework has been described, we will propose a fuzzy treatment of the weights expressing the relative importance of the decision criteria and we will incorporate this treatment into the UW-TOPSIS algorithm. In Sect. 3, a real case study will be presented. We will illustrate the proposed approach ranking a sample of Finnish companies based on their gender equality degree. Finally, in Sect. 4, the main conclusions of the work will be presented.

2 Unweighted TOPSIS with Linguistic Intervals

Technique for Order Preference by Similarity to Ideal Solution (TOPSIS) proposed by Hwang and Yoon (1981) provides a ranking of decision alternatives simultaneously minimizing distance to a positive ideal solution (PIS) and maximizing distance to a negative ideal solution (NIS). The positive ideal solution is composed of the best value of each criterion, and the negative ideal solution is composed of the worst values of the decision criteria. The method provides a cardinal ranking of alternatives, and it is widely used due to its simplicity and nice properties allowing total linear compensation using a single criterion aggregation approach (Behzadian et al., 2012; Chen & Hwang, 1992; Yoon & Hwang, 1995).

As mentioned in the previous section, weights of the criteria in TOPSIS-based approaches are given by the decision-makers a priori in the first steps of the algorithm and maybe objective or subjective depending on the characteristics of the decision problem (Ouenniche et al., 2018; Watröbski et al., 2019). Numerical subjective weights, usually directly established by the decision-maker based on expert knowledge or subjective preferences, are difficult to be uphold, especially in public decision-making. In this work, we propose to handle subjective weights using linguistic terms. In what follows we will illustrate the selected procedure.

2.1 Fuzzy Treatment of Decision Criteria Weights

Let us define a linguistic evaluation scale as the following set

$$l_1 = \{s_\alpha : \alpha \in \{0, 1, \ldots, H\}\}, \tag{1}$$

verifying the following conditions (see Herrera & Martínez, 2000, 2001; Xu, 2004, 2012; Yager, 1995):

i. The set is ordered: $s_\alpha > s_\beta$ if $\alpha > \beta$;
ii. There is a negation operator: $\mathrm{neg}(s_\alpha) = s_\beta$ such that $\beta = H + 1 - \alpha$;

iii. There are max and min operators: $\max(s_\alpha, s_\beta) = s_\alpha$ if $\alpha \geq \beta$, and $\min(s_\alpha, s_\beta) = s_\alpha$ if $\alpha \leq \beta$.

Let us consider the following collection of elements from l_1, $\{s_0, s_2, \ldots, s_p\}$, where $s_0 \leq s_2 \leq \ldots \leq s_p$. Then, following Xu (2004), it is possible to express the collection as an interval $[s_0, s_p]$.

Definition 1 (Xu, 2004). A linguistic interval evaluation scale can be defined as

$$l_2 = \{\tilde{s} = [s_\alpha, s_\beta] : \alpha \leq \beta,\ \alpha, \beta \in \{0, 1, \ldots, H\}\}. \tag{2}$$

The extension to a continuous scale of the previous sets is as follows (Herrera & Martínez, 2000, 2001; Xu, 2004):

$$L_1 = \{s_\alpha : \alpha \in [0, H]\}, \tag{3}$$

$$L_2 = \{\tilde{s} = [s_\alpha, s_\beta] : \alpha, \beta \in [0, H], \alpha \leq \beta\}. \tag{4}$$

Let us now consider a partition of the interval $[0, H]$ into H disjoint subintervals, $\overline{r}_\alpha$, $\alpha = 1, 2, \ldots, H$, such that $\sup \overline{r}_\alpha \leq \inf \overline{r}_{\alpha+1}$, $\alpha = 0, 1, \ldots, H-1$. If we make

$$T(s_\alpha) = \overline{r}_\alpha, \quad \alpha = 0, 1, \ldots, H, \tag{5}$$

we can transform each term of l_1 into an interval contained in $[0, H]$.

Based on Xu (2004), we can define in L_2 the following operation:

$$\lambda \otimes \tilde{s} = \lambda \otimes [s_\alpha, s_\beta] = [s_{\lambda\alpha}, s_{\lambda\beta}], \quad \lambda \in 0, 1, \tag{6}$$

and by its own construction,

$$\lambda \oplus \tilde{s} \in L_2, \quad \lambda \in [0, 1]. \tag{7}$$

2.2 *UW-TOPSIS with Fuzzy Weights*

In what follows we will present the steps of the new algorithm proposed in this paper which does not require the introduction of a priori precise weights. Figure 1 displays the steps in the classical TOPSIS model.

In the UW-TOPSIS approach, the PIS and NIS solutions are determined without consideration of the relative importance of the criteria. In Liern and Pérez-Gladish (2022), weights are introduced as unknown variables in Step 4 when separation

STEP 1. DETERMINE THE DECISION MATRIX D, where the number of criteria is m and the number of alternatives is n, $D = [x_{ij}]_{n \times m}$.

STEP 2. CONSTRUCT THE NORMALIZED DECISION MATRIX. Criteria are expressed in different scaling and therefore a normalizing procedure is necessary to facilitate comparison. Hwang and Yoon (1981) propose a vector normalization,

$$r_{ij} = \frac{x_{ij}}{\sqrt{\sum_{i=1}^{n} (x_{ij})^2}} \in [0,1], \qquad 1 \le i \le n, \qquad 1 \le j \le m. \tag{8}$$

STEP 3. DETERMINE THE WEIGHTED NORMALIZED DECISION MATRIX. It is well known that the weights of the criteria in decision making problems do not have the same mean and not all of them have the same importance. The weighted normalized value v_{ij} is calculated as:

$$v_{ij} = w_j r_{ij}, \qquad 1 \le i \le n, \qquad 1 \le j \le m, \tag{9}$$

where w_j is the weight associated to each criterion.

STEP 4. DETERMINE THE POSITIVE IDEAL (PIS) AND NEGATIVE IDEAL SOLUTIONS (NIS). The positive ideal solution, $A^+ = (v_1^+, \ldots, v_m^+)$, and the negative ideal solution, $A^- = (v_1^-, \ldots, v_m^-)$, are determined as follows:

$$v_j^+ = w_j r_j^+ = \begin{cases} \max\limits_{1 \le i \le n} v_{ij}, & j \in J \\ \min\limits_{1 \le i \le n} v_{ij}, & j \in J' \end{cases} \qquad 1 \le j \le m, \tag{10}$$

$$v_j^- = w_j r_j^- = \begin{cases} \min\limits_{1 \le i \le n} v_{ij}, & j \in J \\ \max\limits_{1 \le i \le n} v_{ij}, & j \in J' \end{cases} \qquad 1 \le j \le m, \tag{11}$$

where J is associated with the criteria that indicate profits or benefits and J' is associated with the criteria that indicate costs or losses.

STEP 5. CALCULATE THE SEPARATION MEASURES. Calculation of the separation of each alternative with respect to the PIS and NIS, respectively:

$$d_i^+ = \left(\sum_{j=1}^{m} (v_{ij} - v_j^+)^2 \right)^{1/2}, \qquad d_i^- = \left(\sum_{j=1}^{m} (v_{ij} - v_j^-)^2 \right)^{1/2}, \qquad 1 \le i \le n. \tag{12}$$

STEP 6. CALCULATE THE RELATIVE PROXIMITY TO THE IDEAL SOLUTION. Calculation of the relative proximity of each alternative to the PIS and NIS using the proximity index.

$$R_i = \frac{d_i^-}{d_i^+ + d_i^-}, \qquad 1 \le i \le n. \tag{13}$$

Th R_i value lies between 0 and 1. If $R_i = 1$, then $A_i = A^+$ and if $R_i = 0$, then $A_i = A^-$. The closer the R_i value is to 1 the higher the priority of the *i-th* alternative.

STEP 7. RANK THE PREFERENCE ORDER. Rank the best alternatives according to R_i in descending order.

Fig. 1 Classical TOPSIS

measures from the PIS and NIS are calculated. Their values are determined in Step 5 solving two groups of mathematical programming problems which maximize and minimize the separation of each alternative to the PIS and NIS, respectively, considering different constraints referred to the values of the weights. These constraints include the classical constraint in TOPSIS approaches which ensures all the weights are positive and sum up one and other constraints imposing lower and upper bounds on the weights. The resulting mathematical programming problems are, due to the nature of their objective, fractional mathematical programming problems. Figure 2 displays the UW-TOPSIS algorithm with weights being unknown variables.

Remark 1 According to Canós and Liern (2008), given the intervals $A = [a_1, a_2]$ and $B = [b_1, b_2]$ contained in $\mathbb{R}$, we will say that A is bigger than B, if and only if

$$A \succ B \Leftrightarrow \begin{cases} k_1 a_1 + k_2 a_2 > k_1 b_1 + k_2 b_2, & k_1 a_1 + k_2 a_2 \neq k_1 b_1 + k_2 b_2 \\ a_1 > b_1, & k_1 a_1 + k_2 a_2 = k_1 b_1 + k_2 b_2 \end{cases}$$

where k_1 and k_2 are two pre-established positive constants. In the context that concerns us, the values k_1 and k_2 inform us about the degree of confidence of the decision-maker that the alternatives are in their best position or on the contrary (Canós & Liern, 2008). When ordering the intervals $\left[R_i^L, R_i^U\right]$, $1 \leq i \leq n$, the relation k_2/k_1 informs us about the importance (or truthfulness) given to the best situation of the alternatives R_i^U regarding of the worst situation R_i^L. In the following examples, since we do not have information that makes us opt for the best or worst situation, we have chosen to give the same importance to both, that is, $k_1 = k_2 = 1$.

Let now us assume a situation in which weights are given by the decision-maker using a linguistic interval evaluation scale as in (2) which in the continuous case takes the form

$$\tilde{W} = \left\{\tilde{w} = [w_\alpha, w_\beta] : \alpha, \beta \in [0, 1],\ \alpha \leq \beta\right\}. \tag{23}$$

To be able to use weights in (23) in the UW-TOPSIS method, it is necessary to establish some conditions:

Definition 2 A vector of weights $\tilde{w} = (\tilde{w}_1, \ldots, \tilde{w}_m) \in \tilde{W}^m$, whose components are intervals with linguistic valuations, is UWL feasible, if it belongs to the following set $\tilde{\Omega}$:

$$\tilde{\Omega} = \left\{\tilde{w} = (\tilde{w}_1, \ldots, \tilde{w}_m) \in \tilde{W}^m,\ \tilde{w_j} = \left[w_{\alpha_j}, w_{\beta_j}\right], \right.$$
$$\left. 1 \leq j \leq m,\ \sum_{j=1}^{m} \alpha_j \leq 1,\ \sum_{j=1}^{m} \beta_j \geq 1 \right\} \tag{24}$$

Given a non-null vector $\tilde{w} = (\tilde{w}_1, \ldots, \tilde{w}_m) \in \tilde{W}^m$, we can obtain a vector of weights UWL feasible if

STEP 1. DETERMINE THE DECISION MATRIX $[x_{ij}]$, $1 \le i \le n$, $1 \le j \le m$, WHERE THE NUMBER OF ALTERNATIVES IS N AND THE NUMBER OF CRITERIA IS M.

STEP 2. CONSTRUCT THE NORMALIZED DECISION MATRIX

$$[r_{ij}],\ r_{ij} \in [0,1],\quad 1 \le i \le n,\ 1 \le j \le m. \tag{14}$$

STEP 3. DETERMINE THE POSITIVE IDEAL $A^+ = (r_1^+, \dots, r_m^+)$ AND THE NEGATIVE IDEAL SOLUTIONS

$A^- = (r_1^-, \dots, r_m^-)$, WHERE

$$r_j^+ = \begin{cases} \max\limits_{1 \le i \le n} r_{ij}, & j \in J \\ \min\limits_{1 \le i \le n} r_{ij}, & j \in J' \end{cases} \quad 1 \le j \le m, \tag{15}$$

$$r_j^- = \begin{cases} \min\limits_{1 \le i \le n} r_{ij}, & j \in J \\ \max\limits_{1 \le i \le n} r_{ij}, & j \in J' \end{cases} \quad 1 \le j \le m, \tag{16}$$

WHERE J IS ASSOCIATED WITH "THE MORE, THE BETTER" CRITERIA AND J' IS ASSOCIATED WITH "THE LESS, THE BETTER" CRITERIA.

STEP 4. LET US CONSIDER $\Omega = \{w = (w_1, \dots, w_m) \in \mathbb{R}^m,\ w_j \in [0,1],\ \sum_{j=1}^m w_j = 1\}$. GIVEN A^+, A^-, WE DEFINE TWO SEPARATION FUNCTIONS,

$$D_i^+ : \Omega \times \mathbb{R}^m \longrightarrow [0,1],\qquad D_i^- : \Omega \times \mathbb{R}^m \longrightarrow [0,1],\quad 1 \le i \le n,$$

GIVEN BY

$$D_i^+(w) = d\big((w_1 r_{i1}, \dots, w_m r_{im}), (w_1 r_1^+, \dots, w_m r_m^+)\big),\quad 1 \le i \le n, \tag{17}$$

$$D_i^-(w) = d\big((w_1 r_{i1}, \dots, w_m r_{im}), (w_1 r_1^-, \dots, w_m r_m^-)\big),\quad 1 \le i \le n, \tag{18}$$

WHERE D IS A DISTANCE FUNCTION IN $\mathbb{R}^m$.

STEP 5. CALCULATE THE FUNCTION OF RELATIVE PROXIMITY TO THE IDEAL SOLUTION, $R_i : \Omega \longrightarrow [0,1]$, $1 \le i \le n$, AS

$$R_i(w) = \frac{D_i^-(w)}{D_i^+(w) + D_i^-(w)},\quad 1 \le i \le n. \tag{19}$$

STEP 6. FOR EACH I, $1 \le i \le n$, WE CALCULATE THE VALUES $R_i^L(w), R_i^U(w)$ SOLVING THE TWO FOLLOWING MATHEMATICAL PROGRAMMING PROBLEMS WHERE DECISION VARIABLES ARE THE CRITERIA WEIGHTS:

$$R_i^L = \text{MIN}\left\{R_i(w), \sum_{j=1}^m w_j = 1,\ l_j \le w_j \le u_j,\ 1 \le j \le m\right\},\ 1 \le i \le n, \tag{20}$$

$$R_i^U = \text{MAX}\left\{R_i(w), \sum_{j=1}^m w_j = 1,\ l_j \le w_j \le u_j,\ 1 \le j \le m\right\},\ 1 \le i \le n, \tag{21}$$

BEING $l_j, u_j \ge 0$ LOWER AND UPPER BOUNDS FOR EACH CRITERION'S WEIGHT. THEN, WE OBTAIN N RELATIVE PROXIMITY INTERVALS,

$$R_i^I = [R_i^L, R_i^U],\quad 1 \le i \le n. \tag{22}$$

STEP 7. WE RANK THE INTERVALS $R_1^I, R_2^I, \dots, R_n^I$ (SEE REMARK 1).

Fig. 2 UW-TOPSIS

Proposition 1 *Given a vector* $\tilde{w} = (\tilde{w}_1, \dots, \tilde{w}_m) \in \tilde{W}^m$, $\widetilde{w_j} = [w_{\alpha_j}, w_{\beta_j}]$, $1 \leq j \leq m$, *with at least one* $\beta_j \neq 0$, *it is possible to construct a vector* $\tilde{w}^* = \left(\tilde{w}_1^*, \dots, \tilde{w}_1^*\right) \in \tilde{\Omega}$, *such that*

$$\widetilde{w_j} = \left[w_{\alpha_j}, w_{\beta_j}\right] \subseteq \tilde{w}_1^* = \left[w_{\alpha_j^*}, w_{\beta_j^*}\right], \quad 1 \leq j \leq m.$$

Proof We will give a constructive demonstration considering the conditions to belong to $\tilde{\Omega}$.

(a) If $\sum_{j=1}^m \alpha_j > 1$, applying (6), we can make

$$w_{\alpha_j^*} = \frac{1}{M \sum_{j=1}^m \alpha_j} w_{\alpha_j}, \quad 1 \leq j \leq m, \ M > 1, \tag{25}$$

verifying

$$w_{\alpha_j^*} \leq w_{\alpha_j}, \quad 1 \leq j \leq m. \tag{26}$$

(b) If $\sum_{j=1}^m \beta_j < 1$, as by hypothesis $\beta_{j_0} \neq 0$, we make

$$w_{\beta_{j_0}^*} = \frac{1}{\beta_{j_0}} w_{\beta_{j_0}} \quad \text{and} \quad w_{\beta_j^*} = w_{\beta_j}, \ j \neq j_0, \tag{27}$$

which verify

$$w_{\beta_j} \leq w_{\alpha_j^*}, \quad 1 \leq j \leq m. \tag{28}$$

Remark 2 Of course, the construction given in (27) and (28) is not the only way to demonstrate Proposition 1. On the other hand, Por otro lado, it is worth highlighting the value $M = 1$ in expression (25) because in this case, $\sum_{j=1}^m \alpha_j^* = 1$ and, as we will see in the next section, this makes it so that when weights are applied to a multi-criteria method (as in the case of TOPSIS), the values $[w_{\alpha_j^*}, w_{\beta_j^*}]$, $1 \leq j \leq m$ cannot be true intervals, but the values $w_{\alpha_j^*}$, $1 \leq j \leq m$.

Steps 1–3 will remain the same than in the UW-TOPSIS algorithm. However, the remaining steps in the algorithm will be transformed as follows:

Step 4. Given a weight UWL feasible $\tilde{w} \in \tilde{\Omega}$, we construct set

$$\Omega_{\tilde{w}} = \left\{ w = (w_1, \dots, w_m) \in W^m, \ \alpha_j \leq w_j \leq \beta_j, \ \sum_{j=1}^m w_j = 1, \ 1 \leq j \leq m \right\}. \tag{29}$$

Given A^+, A^-, we define two separation functions,

$$D_i^+ : \Omega_{\tilde{w}} \times \mathbb{R}^m \to [0, 1], \quad D_i^- : \Omega_{\tilde{w}} \times \mathbb{R}^m \to [0, 1], \quad 1 \leq i \leq n.$$

Given by

$$D_i^+(w) = d\big((w_1 r_{i1}, \ldots, w_m r_{im}), \big(w_1 r_1^+, \ldots, w_m r_m^+\big)\big), \quad 1 \leq i \leq n, \tag{30}$$

$$D_i^-(w) = d\big((w_1 r_{i1}, \ldots, w_m r_{im}), \big(w_1 r_1^-, \ldots, w_m r_m^-\big)\big), \quad 1 \leq i \leq n, \tag{31}$$

where d is a distance function in $\mathbb{R}^m$.

Step 5. Calculate the function of relative proximity to the ideal solution, $R_i : \Omega \to [0, 1]$, $1 \leq i \leq n$, as

$$R_i(w) = \frac{D_i^-(w)}{D_i^+(w) + D_i^-(w)}, \quad 1 \leq i \leq n. \tag{32}$$

Step 6. For each i, $1 \leq i \leq n$, we calculate the values $R_i^L(w)$, $R_i^U(w)$ solving the two following mathematical programming problems where decision variables are the criteria weights:

$$R_i^L = \text{Min}\{R_i(w),\ w \in \Omega_{\tilde{w}}\}, \quad R_i^U = \text{Max}\{R_i(w),\ w \in \Omega_{\tilde{w}}\}. \tag{33}$$

Then, we obtain n relative proximity intervals

$$R_i^I = \big[R_i^L, R_i^U\big], \quad 1 \leq i \leq n. \tag{34}$$

Step 7. We rank the intervals R_1^I, R_2^I, ..., R_n^I (see Remark 1).

Definition 3 We will call diversity and inclusion adequacy index (DIAI) of alternative i

$$\text{DIAI}_i = \frac{R_i^L + R_i^U}{2}, \quad 1 \leq i \leq n. \tag{35}$$

In the next section, we will illustrate our method with a real example in a decision-maker establishing the importance of the diversity and inclusion criteria using linguistic terms. As we will see, these valuations will give rise to a linguistic interval expressing the importance of each criterion. Once these linguistic intervals are obtained, and after verification of the previously described properties, the weights will be integrated in the UW-TOPSIS algorithm, and a set of firms will be assessed in terms of their diversity and inclusion adequacy.

3 Ranking Finnish Companies Based on Their Gender Equality Degree Using UW-TOPSIS with Linguistic Variables

In order to illustrate the proposed assessment method, we will measure the degree of gender diversity and inclusion of a sample of 26 Finnish companies (see Table 1).

Following Equileap, we will assess firms using 19 gender equality criteria organized into four main dimensions: gender balance in leadership and workforce, fair remuneration, policies promoting gender equality, and commitment, transparency

Table 1 Selected Finnish companies

Firm	Sector	Group
F_1	Basic materials	Paper and forest products
F_2	Telecommunications services	Telecommunications services
F_3	Utilities	Electric utilities and IPPs
F_4	Energy	Oil and gas
F_5	Technology	Software and IT services
F_6	Basic materials	Paper and forest products
F_7	Consumer non-cyclicals	Food and drug retailing
F_8	Technology	Communications and networking
F_9	Basic materials	Chemicals
F_{10}	Telecommunications services	Telecommunications services
F_{11}	Basic materials	Paper and forest products
F_{12}	Consumer cyclicals	Automobiles and auto parts
F_{13}	Industrials	Industrial conglomerates
F_{14}	Health care	Pharmaceuticals
F_{15}	Consumer cyclicals	Household goods
F_{16}	Consumer cyclicals	Media and publishing
F_{17}	Industrials	Machinery, tools, heavy vehicles, trains, and ships
F_{18}	Basic materials	Containers and packaging
F_{19}	Basic materials	Metals and mining
F_{20}	Basic materials	Paper and forest products
F_{21}	Industrials	Machinery, tools, heavy vehicles, trains, and ships
F_{22}	Financials	Real estate operations
F_{23}	Industrials	Machinery, tools, heavy vehicles, trains, and ships
F_{24}	Industrials	Machinery, tools, heavy vehicles, trains, and ships
F_{25}	Industrials	Machinery, tools, heavy vehicles, trains, and ships
F_{26}	Financials	Insurance

Source Equileap (2019)

Table 2 Gender diversity criteria

Criteria	Description
C_1	Gender balance in leadership and workforce
C_2	Fair remuneration
C_3	Policies promoting gender equality
C_4	Commitment, transparency, and accountability

Source A detailed description of the decision criteria can be found in Liern and Pérez-Gladish (2022) and Equileap (2019)

and accountability (see Table 2). Table 3 shows the initial decision matrix, with C_i, $i = 1, \ldots, 4$ diversity criteria and $A_j, j = 1, \ldots, 26$ companies (our decision alternatives).

Let us assess the degree of diversity and inclusion of the firms using FUW-TOPSIS. First, we define a linguistic evaluation scale for the relative importance of the decision criteria as the following finite and totally ordered discrete term set composed of five possible values for the linguistic variable representing the weight of criterion *i*. Let us suppose all the criteria weights are described by the same linguistic evaluation scale:

$$l_1 = \{s_0, s_{0.2}, s_{0.4}, s_{0.6}, s_{0.8}, s_1\} \tag{36}$$

being

$s_0 =$ not important
$s_{0.2} =$ slightly important
$s_{0.4} =$ moderately important
$s_{0.6} =$ important
$s_{0.8} =$ very important
$s_1 =$ essential.

Reasoning as in (4) and (5), we obtain the following sets

$$L_1 = \{s_\alpha/\alpha \in [0, 1]\}, \ \ L_2 = \left\{\tilde{s} = \left[s_\alpha, s_\beta\right] : \alpha, \beta \in [0, 1], \ \alpha \leq \beta\right\}. \tag{37}$$

Table 4 displays the importance of the diversity decision criteria weights in linguistic terms.

For weights in Table 4 being UWL feasible (24), we apply Proposition 1, making $M = 2$ in expression (25),

$$W^* = \left\{\left[\frac{1}{2\sum_{k=1}^{T}\alpha_k}s_{\alpha_k}, s_{\beta_k}\right], \ \ 1 \leq k \leq T\right\}. \tag{38}$$

If we apply (38) to the third column in Table 4, we obtain the set of weights expressing the relative importance of the decision criteria

$$W^* = \left\{\left[\frac{1}{3.2}s_{0.6}, s_1\right], \left[\frac{1}{3.2}s_{0.6}, s_1\right], \left[\frac{1}{3.2}s_{0.2}, s_{0.6}\right], \left[\frac{1}{3.2}s_{0.2}, s_{0.4}\right]\right\}.$$

Table 3 Decision matrix

Firm	Sector	C_1	C_2	C_3	C_4
F_1	Basic materials	29.3	16.4	17.5	0.0
F_2	Telecommunications services	32.0	10.9	17.5	2.5
F_3	Utilities	29.3	13.6	17.5	0.0
F_4	Energy	24.0	16.4	15.0	0.0
F_5	Technology	24.0	13.6	17.5	0.0
F_6	Basic materials	29.3	6.8	17.5	0.0
F_7	Consumer non-cyclicals	24.0	13.6	15.0	0.0
F_8	Technology	24.0	8.2	17.5	2.5
F_9	Basic materials	29.3	6.8	15.0	0.0
F_{10}	Telecommunications services	26.7	8.2	15.0	0.0
F_{11}	Basic materials	21.3	9.5	17.5	0.0
F_{12}	Consumer cyclicals	26.7	5.5	15.0	0.0
F_{13}	Industrials	26.7	5.5	15.0	0.0
F_{14}	Health care	21.3	8.2	15.0	0.0
F_{15}	Consumer cyclicals	21.3	8.2	15.0	0.0
F_{16}	Consumer cyclicals	21.3	8.2	12.5	0.0
F_{17}	Industrials	21.3	5.5	15.0	0.0
F_{18}	Basic materials	18.7	5.5	17.5	0.0
F_{19}	Basic materials	16.0	5.5	15.0	0.0
F_{20}	Basic materials	10.7	8.2	17.5	0.0
F_{21}	Industrials	13.3	5.5	17.5	0.0
F_{22}	Financials	16.0	6.8	12.5	0.0
F_{23}	Industrials	13.3	5.5	15.0	0.0
F_{24}	Industrials	13.3	5.5	15.0	0.0
F_{25}	Industrials	13.3	5.5	15.0	0.0
F_{26}	Financials	16.0	5.5	10.0	0.0

Source Equileap (2019)
Note For each firm, F_i, subindex i shows its position in Equileap's ranking

Table 4 Importance of the decision criteria

	Criteria	Linguistic terms	Intervals
C_1	Gender balance in leadership and workforce	[Important, very important]	$[s_{0.6}, s_1]$
C_2	Fair remuneration	[Important, very important]	$[s_{0.6}, s_1]$
C_3	Policies promoting gender equality	[Slightly important, important]	$[s_{0.2}, s_{0.6}]$
C_4	Commitment, transparency, and accountability	[Slightly important, moderately important]	$[s_{0.2}, s_{0.4}]$

Source Own

Table 5 Relative importance of the decision criteria

	Criteria	Linguistic	UW-TOPSIS bounds	
		Interval	l_j	u_j
C_1	Gender balance in leadership and workforce	$[s_{0.1875}, s_1]$	0.1875	1
C_2	Fair remuneration	$[s_{0.1875}, s_1]$	0.1875	1
C_3	Policies promoting gender equality	$[s_{0.0625}, s_{0.6}]$	0.0625	0.6
C_4	Commitment, transparency, and accountability	$[s_{0.0625}, s_{0.4}]$	0.0625	0.4

Source Own

In Table 5, we have displayed the new weights and their use as bounds in the UW-TOPSIS method.

Table 6 shows the obtained scores applying UW-TOPSIS. In the second column, we have displayed the minimum relative proximity value that each firm can obtain given the weights in Table 5. In the third column, we display the maximum possible value, and in last column, we have displayed the average value which we consider the diversity and inclusion index.

Figure 3 displays the obtained results graphically.

Figure 3 shows the worst, best, and average possible results in terms of diversity and inclusion of each firm, given the weights expressed by the decision-maker in linguistic terms using linguistic intervals. Firm F_2 ranks the first one, followed by firm F_8. However, they both present a big amplitude of their relative proximity intervals which means greater ambiguity and imprecision. Position of firm F_{22} is, for instance, more stable, as the amplitude of the relative proximity interval is small. Each position in the ranking, worst and best, has an associated set of weights given in linguistic terms that can be interpreted as weaknesses and strengths of the firms in terms of the diversity and inclusion decision criteria.

4 Conclusions

In this work, we have shown how an extension of TOPSIS can contribute to the assessment and ranking of firms in terms of their diversity adequacy degree. The proposed method allows the ranking of the decision alternatives without a priori determination of a precise weighting scheme. The main contribution of our proposal is the use of linguistic labels transformed into linguistic intervals incorporated into the UW-TOPSIS algorithm to rank a set of decision alternatives. With our method, the relative proximity to the positive ideal solution is optimized for each firm based on

Table 6 Obtained results

Firms	Min R_i	Max R_i	$DIAI_i$
F_1	0.169470863	0.787936101	0.478703482
F_2	0.549559488	0.929459363	0.739509425
F_3	0.149526912	0.676063302	0.412795107
F_4	0.153238539	0.779985161	0.466611850
F_5	0.136896660	0.668261752	0.402579206
F_6	0.100941001	0.648569355	0.374755178
F_7	0.129331318	0.668011039	0.398671179
F_8	0.343894743	0.888739909	0.616317326
F_9	0.100129071	0.648279611	0.374204341
F_{10}	0.099568687	0.602317635	0.350943161
F_{11}	0.100173194	0.445748992	0.272961093
F_{12}	0.082432026	0.577738735	0.330085381
F_{13}	0.082432026	0.577738735	0.330085381
F_{14}	0.079913502	0.434864431	0.257388966
F_{15}	0.079913502	0.434864431	0.257388966
F_{16}	0.070983113	0.434335707	0.252659410
F_{17}	0.057156154	0.417495629	0.237325892
F_{18}	0.046312094	0.450781654	0.248546874
F_{19}	0.031321757	0.340987609	0.186154683
F_{20}	0.041050744	0.451877937	0.246464341
F_{21}	0.024073463	0.432012017	0.228042740
F_{22}	0.041363915	0.224162803	0.132763359
F_{23}	0.019392000	0.330523138	0.174957569
F_{24}	0.019392000	0.330523138	0.174957569
F_{25}	0.019392000	0.330523138	0.174957569
F_{26}	0.028663844	0.218147224	0.123405534

the possible linguistic intervals expressing the criteria weights. As a result, we obtain a relative proximity interval informing the decision-maker about the worst and best possible positions of each firm in the ranking. This could provide a useful information in terms of improvement opportunities for the firms and allows the decision-maker to express certain preferences regarding the decision criteria using linguistic terms.

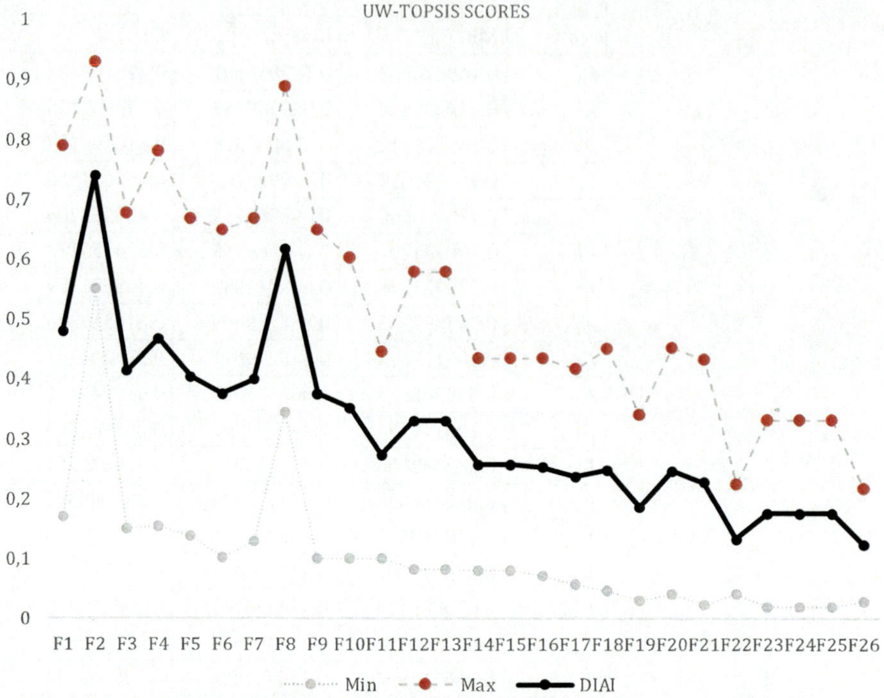

Fig. 3 Results applying UW-TOPSIS with weights displayed in Table 5

Acknowledgements This work has received financial support from the Spanish Ministerio de Ciencia, Innovación y Universidades, project reference number: RTI2018-093541-B-I00.

References

Alemi-Ardakani, M., Milani, A. S., Yannacopoulos, S., & Shokouhi, G. (2016). On the effect of subjective, objective and combinative weighting in multiple criteria decision making: A case study on impact optimization of composites. *Expert Applications, 46*, 426–438.

Barron, F. H., & Barrett, B. E. (1996). Decision quality using ranking attribute weights. *Management Science, 42*, 1515–1525.

Behzadian, M., Otaghsara, S. K., Yazdani, M., & Ignatius, J. (2012). A state-of the-art survey of TOPSIS applications. *Expert Systems with Applications, 39*(7), 13051–13069.

Canós, L., & Liern, V. (2008). Soft computing-based aggregation methods for human resource management. *European Journal of Operational Research, 189*, 669–681.

Chen, S. J., & Hwang, C. L. (1992). *Fuzzy multiple attribute decision making methods and applications* (Vol. 375). Springer-Verlag.

Deng, H., Yeh, C. H., & Willis, R. J. (2000). Inter-company comparison using modified TOPSIS with objective weights. *Computers & Operations Research, 27*, 963–973.

Deszo, C., & Ross, D. (2012). Does female representation in top management improve firm performance? A panel data investigation. *Strategic Management Journal, 33*(9), 1072–1089.

Equileap. (2019). *Gender equality global report & ranking*. https://equileap.org/2019-global-report/. Accessed November 15, 2019.

Eshlaghy, A. T., & Radfar, R. (2006). A new approach for classification of weighting methods. *Management of Innovation and Technology, 2*, 1090–1093.

Herrera, F., & Martínez, L. (2000). A 2-tuple fuzzy linguistic representation model for computing with words. *IEEE Transactions on Fuzzy Systems, 8*, 746–752.

Herrera, F., & Martínez, L. (2001). The 2-tuple linguistic computational model: Advantages of its linguistic description, accuracy and consistency. *International Journal of Uncertainty, Fuzziness and Knowledge-Based Systems, 29*, 33–48.

Hobbs, B. F. (1980). A comparison of weighting methods in power plant sitting. *Decision Science, 11*, 725–737.

Hwang, C. L., & Yoon, K. (1981). *Multiple attribute decision making: Methods and applications: A state-of-the-art survey*. Springer-Verlag.

Liern, V., & Pérez-Gladish, B. (2022). Multiple criteria ranking method based on functional proximity index: Un-weighted TOPSIS. *Annals of Operations Research, 311*, 1099–1121.

Németh, B., Molnár, A., Bozóki, S., Wijaya, K., Inotai, A., Campbell, J. D., & Kaló, Z. (2019). Comparison of weighting methods used in multicriteria decision analysis frameworks in health-care with focus on low- and middle-income countries. *Journal of Comparative Effectiveness Research, 8*(4), 195–204.

Ouenniche, J., Pérez-Gladish, B., & Bouslah, K. (2018). An out-of-sample framework for TOPSIS-based classifiers with application in bankruptcy prediction. *Technological Forecasting and Social Change, 131*, 111–116.

Richard, O., McMillan, A., Chadwick, K., & Dwyer, S. (2003). Employing an innovation strategy in racially diverse workforces. *Group & Organization Management, 28*(1), 107–126.

SHRM Homepage. (2022). *Diversity, equity and inclusion*. https://www.shrm.org/resourcesandtools/hr-topics/pages/diversity-equity-and-inclusion.aspx. Accessed July 10, 2022.

Torchia, M., Calabrò, A., & Huse, M. (2011). Women on corporate boards—From tokenism to critical mass. *Journal of Business Ethics, 102*.

Vafaei, A., Henry, D., Ahmed, K., & Alipour, M. (2020). Board diversity: Female director participation and corporate innovation. *International Journal of Accounting & Information Management, 29*(2), 247–279.

Watröbski, K., Jankiwski, J., Ziemba, P., & Karczmarczyk, A. (2019). Generalised framework for multi-criteria method selection. *Omega, 86*, 107–124.

Whiting, V. (2021). *Review of research literature that provides evidence of the impact of diversity and inclusion in the workplace*. Financial Conduct Authority.

Xu, Z. S. (2004). A method based on linguistic aggregation operators for group decision making with linguistic preference relations. *Information Sciences, 166*, 19–30.

Xu, Z. S. (2012). *Linguistic decision making: Theory and methods*. Springer.

Yager, R. R. (1995). An approach to ordinal decision making. *International Journal of Approximate Reasoning, 12*, 237–261.

Yoon, K. P., & Hwang, C. L. (1995). *Multiple attribute decision making: An introduction*. Sage.

A Multicriteria Analysis Approach to Tourists' Satisfaction with Local Food Consumption

Antonios Tiganis, Stratos Kartsonakis, Efthimia Tsakiridou, and Evangelos Grigoroudis

Abstract In this chapter, we study tourists' satisfaction with local food consumption. Greece is selected as a case study because of the importance of its culinary tradition, while for data collection we interviewed tourists departing from the Thessaloniki Airport "Macedonia". The analysis is based on an extension of the MUSA method. The MUSA method is a multicriteria analysis approach that can collectively measure customers' overall and partial satisfaction, providing a series of results that can identify the strengths and weaknesses of customer perceptions. The results show that tourists are highly satisfied by consuming local food. The most critical local food attributes are taste, safety, aroma, authenticity, appearance, and connection to Greek culture. These attributes are the competitive advantages of local food. On the other hand, healthiness, quality, cost, and package could be perceived as potential threats to tourists' satisfaction. Tourists appear indifferent towards the enhancement of the local economy.

Keywords Tourist satisfaction · Local foods · MUSA method · Tourism food consumption · Multicriteria analysis

A. Tiganis
Aarhus University, Fuglesangs Allé 4, 8210 Aarhus, Denmark
e-mail: atiganis2@huskers.unl.edu

S. Kartsonakis · E. Grigoroudis (✉)
Technical University of Crete, University Campus, 73100 Chania, Greece
e-mail: vangelis@ergasya.tuc.gr

S. Kartsonakis
e-mail: ekartsonakis@isc.tuc.gr

E. Tsakiridou
Aristotle University of Thessaloniki, 54124 Thessaloniki, Greece
e-mail: efitsaki@agro.auth.gr

C. Zopounidis et al. (eds.), *Operational Research Methods in Business, Finance and Economics*, Lecture Notes in Operations Research,
https://doi.org/10.1007/978-3-031-31241-0_5

1 Introduction

Local food covers multiple roles for tourism experience and tourists' wellbeing (Sánchez-Cañizares & López-Guzmán, 2012; Tikkanen, 2007). It can be a primary reason for choosing a destination (Björk & Kauppinen-Räisänen, 2016), as it serves as a central holiday experience by reflecting national and local traits and connecting tourists with the destination's culture (Björk & Kauppinen-Räisänen, 2016). The importance of local food consumption is also related to a sustainable tourism experience by connecting with the destination's culture and people (Sims, 2009). In general, food plays a big part of overall tourists' expenditure (Kim et al., 2009), while tourists show a strong interest in purchasing local food (Sanchez-Cañizares & Castillo-Canalejo, 2015; Sánchez-Cañizares & López-Guzmán, 2012). Cross-cultural research also reveals that tourists are willing to pay more for local food (Akdag et al., 2018; Sanchez-Cañizares & Castillo-Canalejo, 2015).

Previous research efforts have studied the motives that drive tourists to consume local foods in the host destination. Sensory traits, authenticity quest, health concerns, cultural connection, and visual representation are solid motivators for local food consumption (Chang & Mak, 2018; Cohen & Avieli, 2004; Kim et al., 2009; Mak et al., 2012, 2017). Furthermore, there is some evidence for which food attributes tourists perceive as important. Taste is the most important attribute, followed by quality, local origin, and authenticity (Altintzoglou et al., 2016).

On the other hand, there is little research about the drivers of tourists' satisfaction with local gastronomic experiences. Relevant literature mainly focused on the dimensions of tourists' gastronomic satisfaction, the effect of foods' perceived image and value, and its impact on revisit intentions. Food quality, price, variety, convenience, cultural aspects, and appearance affect tourists' satisfaction by gastronomic experiences and food consumption (Chi & Qu, 2008, 2009; Peštek & Činjarević, 2014).

Therefore, it is not clear yet how food attributes affect tourists' satisfaction. This study applies a multicriteria analysis approach to tourist satisfaction, aiming to evaluate the importance of food attributes. The primary research aim of the study is to investigate tourists' satisfaction with local food consumption.

The importance of customer satisfaction is well explained in the relevant literature. It is a predictor of consumers' post-purchase behavior (Grigoroudis & Siskos, 2010), while in the tourist literature, it affects their loyalty to a destination (Hammami et al., 2018; Kim et al., 2011). Estimating the weights or importance of food attributes may help policymakers to develop actions or strategies that can enhance customers' satisfaction.

There is a rich literature discussing the relationships between food consumption, tourist satisfaction, and behavioral intentions. There might be different linkages between the aforementioned variables and alternative mediators. For example, local food consumption motivations are linked with tourists' satisfaction (Perçin et al., 2021), local food experiences may significantly affect tourists' behavioral intention

(Ghanem, 2019), while tourist's involvement can serve as a mediator in the food consumption-satisfaction relationship (Rehman et al., 2022).

In this context, we apply the MUSA (Multicriteria Satisfaction Analysis) method. It is a preference disaggregation technique based on ordinal regression analysis. The MUSA method measures and analyzes satisfaction (consumers' satisfaction, employees' satisfaction, customers' satisfaction), and its results can estimate the importance (weight) of each satisfaction criterion (Grigoroudis & Siskos, 2002). Moreover, MUSA can estimate performance indices that show the average satisfaction level of customers. Based on these results, the MUSA method can generate an action diagram, a matrix similar to SWOT analysis, identifying the strong and weak parts of tourists' satisfaction. It should be also mentioned that the LP formulation of the MUSA method allows the consideration of additional constraints with special properties of the assessed model variables. Under this context, an extension of the MUSA method is applied to the examined problem. A detailed description of the method is presented in the next section.

2 MUSA Method

2.1 Basic Model

The MUSA method, developed by Grigoroudis and Siskos (2002), is the primary research methodology of the presented study. The method aims at achieving the maximum consistency between a collective value function Y^* and a set of partial value functions X_i^*. Partial value functions X_i^* are referring to consumer satisfaction on a specific attribute, while Y^* refers to the overall consumer satisfaction. Using a double-error variable, the ordinal regression equation has the following form:

$$\hat{Y}^* = \sum_{i=1}^{n} b_i X_i^* - \sigma^+ + \sigma^- \tag{1}$$

where $\hat{Y}^*$ is the estimation of the global value function Y^*, n is the number of criteria used in the analysis, b_i is the weight of the i-th criterion with $\sum_{i=1}^{n} b_i = 1$, while σ^+ and σ^- are the overestimation and underestimation errors, respectively.

Both global and partial functions, Y^* and X_i^*, are monotonic and normalized in the interval [0, 100]. To assure monotonicity, the MUSA method uses the following transformation equations:

$$\begin{cases} z_m = y^{*m+1} - y^{*m} \; m = 1, 2, \ldots, \alpha - 1 \\ w_{ik} = b_i x_i^{*k+1} - b_i x_i^{*k} \; k = 1, 2 \ldots, \alpha_i - 1, \; i = 1, 2, \ldots, n \end{cases} \tag{2}$$

where α and α_i are the number of levels of the global and partial value functions, y^{*m} is the value of the y^m overall satisfaction level, and x_i^{*k} is the value of the x_i^k partial satisfaction level.

Using linear programming, the optimization problem can be written as follows:

$$[\min]F = \sum_{j=1}^{M} (\sigma_j^+ + \sigma_j^-)$$

subject to

$$\sum_{i=1}^{n}\sum_{k=1}^{x_i^j-1} w_{ik} - \sum_{m=1}^{y^j-1} z_m - \sigma_j^+ + \sigma_j^- = 0 \quad \text{for } j = 1, 2, \ldots, M$$

$$\sum_{m=1}^{\alpha-1} z_m = 100$$

$$\sum_{i=1}^{n}\sum_{k=1}^{\alpha_i-1} w_{ik} = 100$$

$$z_m, w_{ik}, \sigma_j^+, \sigma_j^- \geq 0 \;\; \forall \; i, j, k, m \tag{3}$$

where M is the number of customers and y^j, x_i^j are overall and partial satisfaction (on the i-th criterion) of the j-th customer using the ordinal scales Y and X_i.

Assuming strictly increasing value functions, the previous LP may be re-written as follows:

$$[\min]F = \sum_{j=1}^{M} (\sigma_j^+ + \sigma_j^-)$$

subject to

$$\sum_{i=1}^{n}\sum_{k=1}^{x_i^j-1} w'_{ik} - \sum_{m=1}^{y^j-1} z'_m - \sigma_j^+ + \sigma_j^- = \gamma(y^j - 1) - \sum_{i=1}^{n} \gamma_i (x_i^j - 1) \quad \text{for } j = 1, 2, \ldots, M$$

$$\sum_{m=1}^{\alpha-1} z'_m = 100 - \gamma(\alpha - 1)$$

$$\sum_{i=1}^{n}\sum_{k=1}^{\alpha_i-1} w'_{ik} = 100 - \sum_{i=1}^{n} \gamma_i(\alpha_i - 1)$$

$$z'_m, w'_{ik}, \sigma_j^+, \sigma_j^- \geq 0 \;\; \forall \; i, j, k, m \tag{4}$$

where γ and γ_i are the preference thresholds for the value functions Y^* and X_i^*, respectively (with $\gamma, \gamma_i \geq 0$) and z'_m, w'_{ik} are the new decision variables with $z'_m = z_m - \gamma$ and $w'_{ik} = w_{ik} - \gamma_i$.

The MUSA method includes a post-optimality analysis step in order to analyze model stability. During post-optimality, the existence of multiple or near-optimal solutions is investigated through the following linear programs:

$$\begin{aligned}
&[\max]F' = \sum_{k=1}^{\alpha_i - 1} w_{ik} \text{ for } i = 1, 2, \dots, n \\
&\text{subject to} \\
&F \leq F^* + \varepsilon \\
&\text{All the constraints of LP(3) or LP(4)}
\end{aligned} \tag{5}$$

where F^* is the optimal value of the objective function F of LP (3) or LP (4) and ε is a small number. The final solution is estimated as the average of the solutions given by the previous n LPs (5).

2.2 Results

Based on the previous modeling approach, the MUSA method estimates the global and partial value functions Y^* and X_i^*, respectively, as follows:

$$y^{*m} = \sum_{t=1}^{m-1} z_t \quad \text{for } m = 2, 3, \dots, \alpha \tag{6}$$

$$x_i^{*k} = 100 \frac{\sum_{t=1}^{k-1} w_{it}}{\sum_{t=1}^{\alpha_i - 1} w_{it}} \quad \text{for } i = 1, 2, \dots, n, \quad k = 2, 3, \dots, \alpha_i - 1 \tag{7}$$

The estimated value functions show the real value, in a normalized interval [0,100], that customers give for each level of the global or marginal ordinal satisfaction scale. The form of these functions indicates the customers' degree of demanding, i.e., demanding customers (convex value function), non-demanding customers (concave value function), and neutral customers (linear form of value function). The MUSA method assumes that Y^* and X_i^* are monotonic, nondecreasing, discrete (piecewise linear) functions.

On the other hand, the satisfaction criteria weights represent the relative importance of the assessed satisfaction dimensions. Based on the model variables of the previous sections, the weights are calculated using the following formula:

$$b_i = \frac{\sum_{t=1}^{\alpha_i - 1} w_{it}}{100} \quad \text{for } i = 1, 2, \dots, n \tag{8}$$

The MUSA method assesses also a set of performance indicators in order to estimate the satisfaction level both globally and per satisfaction criterion. The average global and partial satisfaction indices, S and S_i, respectively, are given by the following formulas:

$$S = \frac{1}{100} \sum_{m=1}^{\alpha} p^m y^{*m} \tag{9}$$

$$S_i = \frac{1}{100} \sum_{k=1}^{\alpha_i} p_i^k x_i^{*k} \quad \text{for } i = 1, 2, \ldots, n \tag{10}$$

where p^m and p_i^k are the frequencies of customers belonging to the y^m and x_i^k satisfaction levels, respectively.

As already noted, the shape of the estimated value functions may indicate the demanding level of customers. In this context, the MUSA method assesses the global and partial demanding indices, D and D_i, respectively, as follows:

$$D = \frac{\sum_{m=1}^{\alpha-1} \left(\frac{100(m-1)}{\alpha-1} - y^{*m} \right)}{100 \sum_{m=1}^{\alpha-1} \frac{m-1}{\alpha-1}} \quad \text{for } \alpha > 2 \tag{11}$$

$$D_i = \frac{\sum_{k=1}^{\alpha_i-1} \left(\frac{100(k-1)}{\alpha_i-1} - x_i^{*k} \right)}{100 \sum_{k=1}^{\alpha_i-1} \frac{k-1}{\alpha_i-1}} \quad \text{for } \alpha_i > 2,\ i = 1, 2, \ldots, n \tag{12}$$

These demanding indices represent the average deviation of the estimated value curves from a "normal" (linear) function. They are normalized in $[-1, +1]$, so customers appear demanding if $D \approx 1$ or $D_i \approx 1$, non-demanding if $D \approx -1$ or $D_i \approx -1$, and neutral if $D \approx 0$ or $D_i \approx 0$.

Finally, the MUSA method can generate a series of action diagrams that indicate customers' strong and weak points by combining weights and average satisfaction indices. These diagrams are similar to a SWOT analysis and result in four quadrants as shown in Fig. 1: status quo, leverage opportunity, transfer resources, and action opportunity [see (Grigoroudis & Siskos, 2010)].

- *Status quo (low performance and low importance)*: Generally, no action is required, given that these satisfaction dimensions are not considered as important by the customers.
- *Leverage opportunity (high performance/high importance)*: This area can be used as advantage against competition. In several cases, these satisfaction dimensions are the most important reasons why customers have purchased the product/service under study.
- *Transfer resources (high performance/low importance)*: Regarding the particular satisfaction dimension, company's resources may be better used elsewhere (i.e. improvement of satisfaction dimensions located in the action opportunity quadrant).

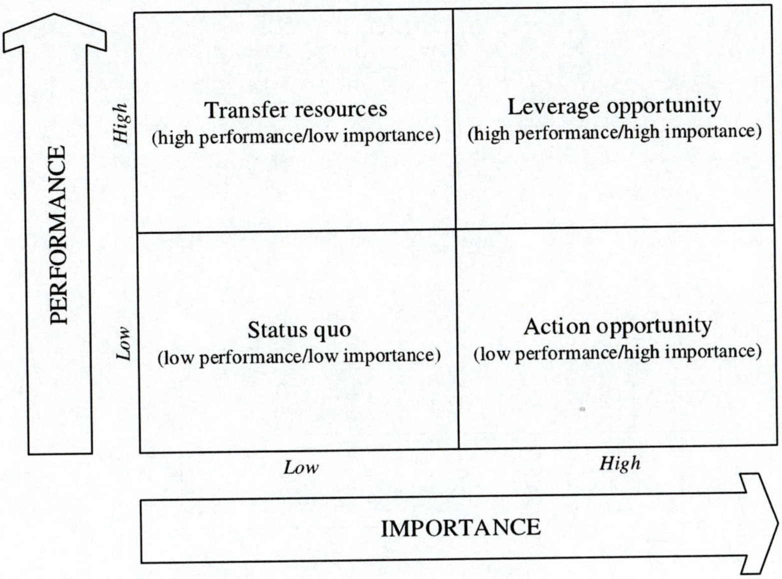

Fig. 1 Action diagram (Grigoroudis & Siskos, 2010)

- *Action opportunity (low performance/high importance)*: These are the criteria that need attention; improvement efforts should be focused on these, in order to increase the global customer satisfaction level.

2.3 *Extension of the MUSA Method*

The basic LP formulation of the MUSA method gives the ability to introduce additional constraints that are able to enhance the stability of the provided results. Grigoroudis and Siskos (2010) examined the introduction of additional constraints regarding the assessed average indices. More specifically, a linkage between global and partial average satisfaction indices may be assumed (the same applies for the average demanding indices) as these considered the main performance indices of the business organizations. Hence, the global average satisfaction S is assessed as a weighted sum of the partial satisfaction S_i:

$$S = \sum_{i=1}^{n} b_i S_i \Leftrightarrow \sum_{m=1}^{\alpha} p^m y^{*m} = \sum_{i=1}^{n} b_i \sum_{k=1}^{\alpha_i - 1} p_i^k x_i^{*k} \tag{13}$$

The previous equation can be re-written using the main variables of LP (3) as follows:

$$\sum_{m=2}^{\alpha} p^m \sum_{t=1}^{m-1} z_t = \sum_{i=1}^{n} \sum_{k=2}^{\alpha_i} p_i^k \sum_{t=1}^{k-1} w_{it} \tag{14}$$

Similarly, a weighted sum formula may be assumed for the average demanding indices:

$$D = \sum_{i=1}^{n} b_i D_i \tag{15}$$

or equivalently:

$$\frac{\sum_{m=1}^{\alpha-1} \left[100(m-1) - (\alpha-1) \sum_{t=1}^{m-1} z_t\right]}{\alpha(\alpha-1)} = \sum_{i=1}^{n} \frac{\sum_{k=1}^{\alpha_i-1} \left[(k-1) \sum_{t=1}^{\alpha_i-1} w_{it} - (\alpha_i - 1) \sum_{t=1}^{k-1} w_{it}\right]}{\alpha_i(\alpha_i-1)} \tag{16}$$

It should be noted that formulas (14) and (16) may be also used in the case of strictly increasing value functions, substituting $z_m = z'_m + \gamma$ and $w_{ik} = w'_{ik} + \gamma_i$.

The previous additional properties for the average satisfaction and demanding indices may be inserted as new constraints in the basic LP formulation. This extension of the MUSA method, proposed by Grigoroudis and Siskos (2010), may provide more robust results (Grigoroudis & Politis, 2015). The applied approach, in the case of the generalized MUSA method (strictly increasing value functions) consists of the following three steps:

Step 1

Solve LP (4).

Step 2

Solve the following LP:

$$[\min]\Phi = \sum_{j=1}^{M} \left[(s_j^+ + s_j^-) + (d_j^+ + d_j^-)\right]$$

subject to

$$\sum_{i=1}^{n} \sum_{k=2}^{\alpha_i} p_i^k \sum_{t=1}^{k-1} w'_{it} - \sum_{m=2}^{\alpha} p^m \sum_{t=1}^{m-1} z'_m - s_j^+ + s_j^- = \gamma \sum_{m=2}^{\alpha} (m-1) p^m - \sum_{i=1}^{n} \gamma_i \sum_{k=2}^{\alpha_i} (k-1) p_i^k$$

$$\sum_{i=1}^{n} \frac{\sum_{k=1}^{\alpha_i-1}\left[\left[1-\gamma_i(\alpha_i-1)(k-1)\right]\sum_{t=1}^{\alpha_i-1} w'_{it} - (\alpha_i-1)\sum_{t=1}^{k-1} w'_{it}\right]}{\alpha_i(\alpha_i-1)}$$

$$-\frac{\sum_{m=1}^{\alpha-1}\left[(100-\gamma(\alpha-1))(m-1)-(\alpha-1)\sum_{t=1}^{m-1} z'_t\right]}{\alpha(\alpha-1)} - d_j^+ + d_j^- = 0$$

$$F \leq F^* + \varepsilon$$

all the constraints of the LP of step 1 (17)

where F^* is the optimal value of the objective function of the LP (4) (step 1), ε is a small number, s_j^+ and s_j^- are the overestimation and underestimation errors, respectively, regarding the average satisfaction indices constraint, and d_j^+, d_j^- are the overestimation and underestimation errors, respectively, regarding the average demanding indices constraint.

Step 3

The final step refers to the stability analysis based on the MUSA III method (Grigoroudis & Siskos, 2010) where the following LP is solved:

$$[\max]F' = z_m \text{ or } [\max]F' = w_{ik}\ \forall i, k, m$$

subject to

$$F \leq F^* + \varepsilon_1$$

$$\Phi \leq \Phi^* + \varepsilon_2$$

all the constraints of steps 1-2 (18)

where Φ^* is the optimal value of the objective function of the LP (17) (step 2) and ε_1, ε_2 are small numbers. The final solution is calculated as the average of the optimal solutions of the previous LPs.

3 Research Design

3.1 Satisfaction Criteria

The assessment of satisfaction criteria in this study is based on previous research efforts that identify which food attributes are considered significant factors for local food consumption and customers' satisfaction.

Like taste and aroma, sensory traits have been identified as essential motivators for food consumption during the holiday (Kim et al., 2009; Mak et al., 2012). Furthermore, the taste is the most vital motivational factor for local food consumption (Altintzoglou et al., 2016).

Quality is one of the essential food attributes during the holiday (Altintzoglou et al., 2016). Quality is also determinant for tourists' gastronomic satisfaction (Akdag et al., 2018; Peštek & Činjarević, 2014). Outside the tourist context, quality concerns are among the strongest drivers for local food consumption (Stephenson & Lev, 2004).

Tourists have been described as authenticity seekers (Cohen & Avieli, 2004). In their quest for authentic experiences, they perceive local cuisine as a conceptual part of place and culture (Henderson, 2009; Sims, 2009). Food authenticity is an essential dimension of the eating experience (Björk & Kauppinen-Räisänen, 2014). Thus, local food consumption is a cultural experience (Wang et al., 2016), allowing tourists to get familiar with the place and cover interpersonal needs (López-Guzmán et al., 2017). Authenticity is a dominant food attribute for tourists (Altintzoglou et al., 2016) and a strong motivator (Kim et al., 2009; Mak et al., 2012).

Health and safety concerns regarding ethnic food consumption were portrayed in various researches (Cohen & Avieli, 2004; Kim & Eves, 2012; Kim et al., 2009; Mak et al., 2012). Healthiness and nutrition are also among the dimensions of the local cuisine image (Peštek & Činjarević, 2014) and among the motivational factors for consuming local foods during the holiday (Kim & Eves, 2012; Kim et al., 2009).

There is a social dimension to local food consumption by tourists as this is beneficial for local societies and economies and tourists (Sims, 2009). When tourists consume local foods through alternative networks, they enhance the local community's sustainability, while these networks are being empowered by consumers who prefer local products (Sims, 2009).

Local cuisine is perceived as a predictor of authenticity by tourists (Cohen & Avieli, 2004), and the eating culture is reflecting national traits through local and national dishes (Björk & Kauppinen-Räisänen, 2016) as food acts as a medium of interaction between humans and places (Ellis et al., 2018). Tourists are seeking to be connected to a host's country culture through local foods (Cohen & Avieli, 2004; Ellis et al., 2018; Tikkanen, 2007) and consider local food culture as an essential dimension of eating experiences (Björk & Kauppinen-Räisänen, 2014; Kim & Eves, 2012). Gastronomic satisfaction is affected by foods' traditional and cultural aspects (Akdag et al., 2018; Peštek & Činjarević, 2014).

Foods' visual image is important both as a motivator for local food consumption (Kim et al., 2009; Mak et al., 2012) and as a satisfaction indicator (Peštek & Činjarević, 2014). The importance of food aesthetics is also recognized as part of the eating experience (Björk & Kauppinen-Räisänen, 2014).

The importance of price is essential for local foods' purchase. Peštek and Činjarević (2014) suggest that price is crucial for local food consumption during holidays. Price is a necessary predictor for purchase intentions (Ahmad et al., 2019), while it can be also a dimension of gastronomic image (Chang & Mak, 2018).

Package is also a vital attribute affecting the purchase of various food products (Endrizzi et al., 2015; Grunert, 1997; Koutsimanis et al., 2012). In the tourism context, package as an extrinsic attribute can be associated with the importance of visual appearance, as indicated in the relevant literature (Kim et al., 2009; Mak et al., 2012).

Based on the aforementioned research and studies, the following satisfaction criteria have been chosen to evaluate local food consumption by tourists:

1. Taste
2. Healthiness
3. Safety
4. Aroma
5. Authenticity
6. Quality
7. Cost/Price
8. Appearance
9. Package
10. Connection to local culture
11. Enhancement to local economy.

3.2 Questionnaire Development

A structured questionnaire has been developed based on the previous satisfaction criteria, and it has been translated into English, German and Russian through verified translators. The questionnaire uses five-point Likert scale questions regarding food consumption evaluations. To investigate tourists' global satisfaction with local food consumption, we asked respondents to state their level of agreement with the following statement: "During the holiday I was satisfied with Greek food consumption". Respondents could state their level of agreement by choosing between the following ordinal scale: Strongly disagree—Disagree—Neither agree nor disagree—Agree—Strongly agree.

Respondents were also asked to evaluate the following attributes of Greek foods: (1) Taste, (2) Healthiness, (3) Safety, (4) Aroma, (5) Authenticity, (6) Quality, (7) Cost/Price, (8) Appearance, (9) Package, (10) Connection to local culture, and (11) Enhancement to the local economy.

Finally, the questionnaire covered some demographic characteristics of tourists, such as gender, age, education, income, and nationality.

3.3 Participants and Sampling

Greece was selected as a case study as its culinary tradition is a vital aspect of choosing Greece as a host destination (Triantafillidou et al., 2019). Positioned at the armpit of the Mediterranean Sea, having suitable soil and climatic conditions for agriculture, being a civilization melting pot for thousands of years, and obtaining a continuous tradition through the centuries are major factors in Greece for the existence of a very competitive and qualified food sector and cuisine. The significance of Greek

Table 1 Socio-demographic characteristics of the sample

Variable	Values	Frequency (% percentage)
Gender	Male	141 (45.3)
	Female	170 (54.7)
Education	Ph.D./Master	151 (48.6)
	Bachelor	103 (33.1)
	Primary/Secondary	57 (18.3)
Household size	1 member	53 (17.0)
	2 members	111 (35.7)
	3 members	55 (17.7)
	4 members	58 (18.7)
	More than 4 members	34 (10.9)
Nationality	Germany	111 (35.7)
	Russia	30 (9.7)
	United Kingdom	41 (13.2)
	Others	129 (41.4)

cuisine is also derived from its connection to the Mediterranean Diet, a part of Human Culture and Intangible Cultural Heritage of UNESCO (Medina, 2009).

The questionnaire was distributed to foreign tourists at the "Macedonia" Airport of Thessaloniki, Central Macedonia, from July 2018 to September 2018. Respondents were tourists who were departing from Greece. For data collection, convenience sampling technique was used. Convenience sampling is used very frequently in tourism research, as it is challenging to apply other techniques. The demographic characteristics of the sample are presented in Table 1.

Most respondents were females (55.2%) and university-level educated (86.3%). The most significant part of the sample had a monthly income greater than 2000 euros, and the biggest nationality category is Germany. In addition, the average age of respondents is 38.83 years (with a standard deviation of 0.81). Overall, the respondents' profile is a German woman with a monthly payment of over 2000 euros and a tertiary education degree.

4 Results

For the analysis, univariate, bivariate, and multivariate methodologies were utilized. Descriptive statistics were used, through STATA 16.0, in order to analyze the demographic traits of the sample. Chi-square and ANOVA, through STATA 16.0, were used to trace the effect of socio-demographic variables to the level of tourists' satisfaction. An extension of the MUSA method was used to analyze customer satisfaction.

Table 2 Satisfaction criteria frequencies (in % percentage*)

	SD	D	NAND	A	SA
Greek foods are tasty	0.00	0.32	1.93	48.55	49.20
Greek foods are healthy	0.32	5.79	22.19	45.66	26.05
Greek foods are safe	0.00	0.96	15.43	55.31	28.30
Greek foods have a nice aroma	0.00	0.96	10.61	53.70	34.73
Greek foods are authentic	0.00	0.32	19.94	46.30	33.44
Greek foods have better quality	0.32	2.89	34.41	40.51	21.86
Greek foods are expensive**	2.25	7.72	31.83	44.05	14.15
Greek foods have a nice appearance	0.00	2.25	22.19	59.16	16.40
Greek foods have a nice package	2.89	12.54	50.16	27.65	6.75
Greek foods are connected to Greek culture	0.00	1.29	20.90	52.73	25.08
Greek foods are enhancing Greek economy	0.32	3.22	39.23	40.51	16.72

*SD: Strongly Disagree; D: Disagree; NAND: Neither Agree Nor Disagree; A: Agree; SA: Strongly Agree
** Reversely coded

4.1 Univariate Analysis

The majority of the sample strongly agrees ($n = 149/47.91\%$) that they are satisfied by local food. Another major part of the respondents stated that they agree to the satisfaction statement ($n = 145/46.62\%$). A small amount of the sample stated that they neither agree nor disagree that they are satisfied with local food consumption ($n = 17/5.47\%$). There are no tourists who disagreed with the satisfaction statement. The results of the food evaluations on the detailed satisfaction criteria are presented in Table 2.

4.2 Bivariate Analysis

The effect of the respondents' socio-demographic traits on the level of their overall satisfaction is presented in Table 3. Gender, household size, income, and age are significant for tourists' satisfaction. Females, respondents who belong to a two-member household, earning monthly more than 3000 euros and with an average age of 39 years old, demonstrate the highest level of agreement with the statement that they are satisfied with local food.

Table 3 Socio-demographic effects on tourists' satisfaction

	During holiday I am satisfied with local food consumption*			Chi-square	*p*-value
	NAND	A	SA		
Gender				6.5653	0.038
Male	70.59	42.07	38.26		
Female	29.41	57.93	61.74		
Education				6.4947	0.165
Ph.D./Master	29.41	55.17	44.30		
Bachelor	47.06	27.59	36.91		
Primary/Secondary	23.53	17.24	18.79		
Household size				16.1440	0.040
1 member	5.88	21.38	14.09		
2 members	11.76	35.17	38.93		
3 members	17.65	16.55	18.79		
4 members	35.29	15.86	19.46		
More than 4 members	29.41	11.03	8.72		
Monthly income				10.8705	0.092
Less than 1,000 euro	35.29	13.79	13.42		
1000–2000 euro	17.65	26.90	20.81		
2000–3000 euro	35.29	28.28	26.85		
More than 3000 euro	11.76	31.03	38.93		
Nationality				2.6717	0.849
Germany	35.29	34.48	36.91		
Russia	23.53	15.86	14.77		
United Kingdom	11.76	13.10	8.72		
Others	29.41	36.55	39.60		

*NAND: Neither Agree Nor Disagree; A: Agree; SA: Strongly Agree

4.3 Multivariate Analysis

The main results of the MUSA method are presented in Table 4. The Average Fitting and Stability Indices of the MUSA method is 89.16% and 74.93%, respectively. These results show that the analyzed customer data are sufficient, and the results of the applied method are highly representative.

Taste is the criterion that has the highest importance (18.9%), which is more than double compared to the weights of the other attributes. Taste is followed by aroma (9.2%), safety (8.3%), authenticity (8.2%), appearance (8.1%) and connection to the Greek culture (8%). The criteria with the lowest weights but with small difference

Table 4 Results of the MUSA method

Criteria	Weight (%)	Average satisfaction index [0,1]	Average demanding index [−1,+1]
Taste	18.9	0.945	−0.560
Healthiness	7.9	0.741	−0.036
Safety	8.3	0.799	−0.061
Aroma	9.2	0.841	−0.147
Authenticity	8.2	0.801	−0.055
Quality	7.8	0.715	−0.031
Cost/Price	7.8	0.664	−0.033
Appearance	8.1	0.747	−0.056
Package	7.6	0.568	−0.023
Connection to Greek culture	8.0	0.771	−0.044
Enhancement to Greek economy	7.7	0.689	−0.030
Overall satisfaction	–	0.842	0.040

compared to the others are package (7%), and enhancement to the Greek economy (7.72%).

Overall, tourists appear quite satisfied since the global average satisfaction index is almost 0.85. The criterion with the highest average satisfaction index is taste (0.945), followed by aroma (0.841), authenticity (0.801), safety (0.799), connection to Greek Culture (0.771) and appearance (0.747). The criteria with the smallest average satisfaction indices are package (0.568), cost/price (0.664), and enhancement to Greek economy (0.689). It can be noticed that criteria with the highest (lowest) performance indices have at the same time the highest (lowest) weights.

The estimated value functions are presented in Fig. 2. They appear to have a rather linear form, revealing that tourists have a neutral demanding level. The average demanding indices further confirm this finding. Both the global and the partial demanding indices are close to zero, showing that the higher satisfaction tourists in Greece express towards local food, the higher the percentage of their fulfilled expectations.

The action diagram is presented in Fig. 3 and can be used to identify the strengths and weaknesses of local food consumption by tourists. The Leverage Opportunity quadrant contains two criteria, i.e., taste and aroma, which have both high-performance indices and high weights, and thus, they are considered as the competitive advantage of local foods. On the other hand, safety, authenticity, and connection the Greek culture belong to the Transfer Resources quadrant. Despite the high performance of these criteria, their impact to tourist's satisfaction is low.

Furthermore, six criteria are located in the Status Quo quadrant: healthiness, appearance, quality, enhancement to the Greek economy, cost/price, and package. These criteria appear to have low performance and low importance and, although

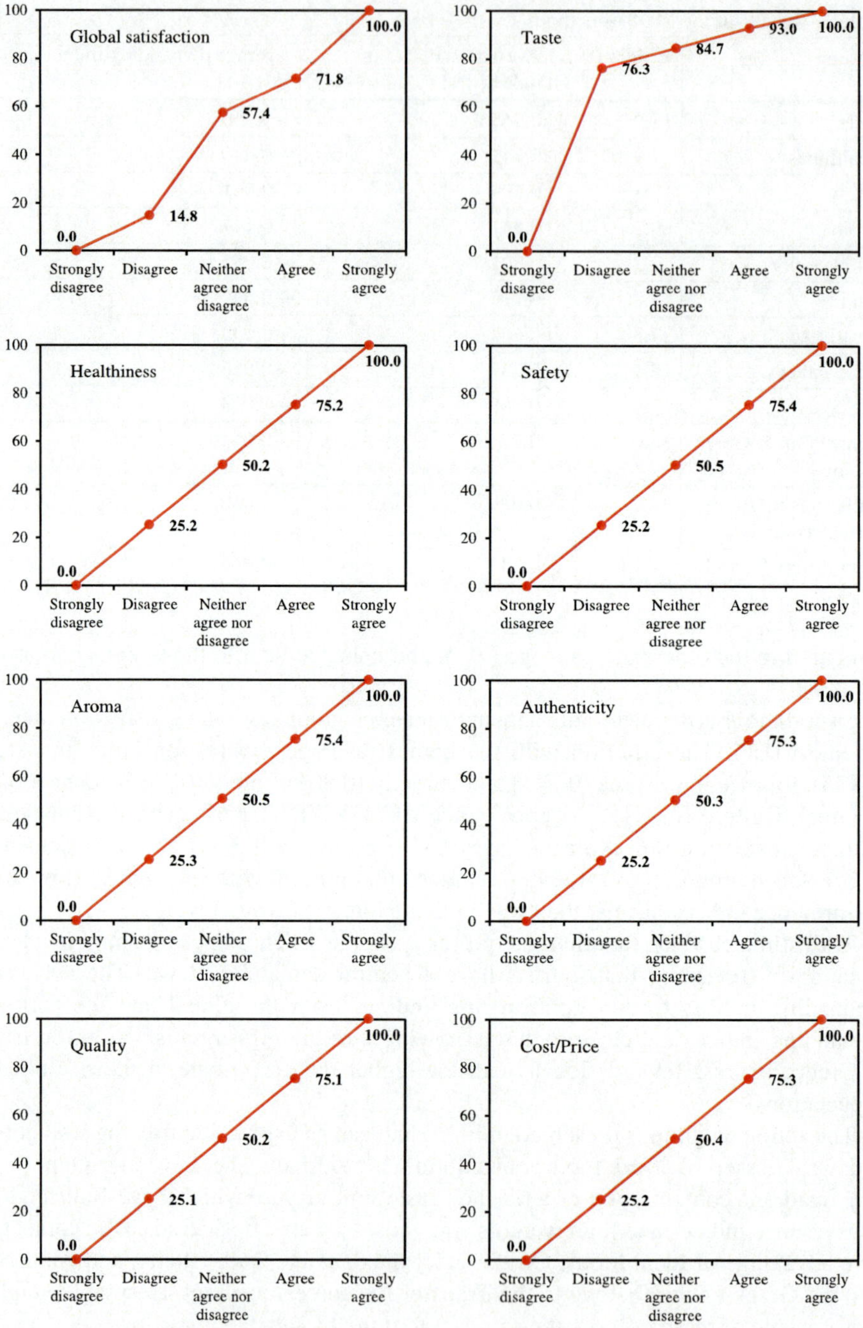

Fig. 2 Estimated value functions

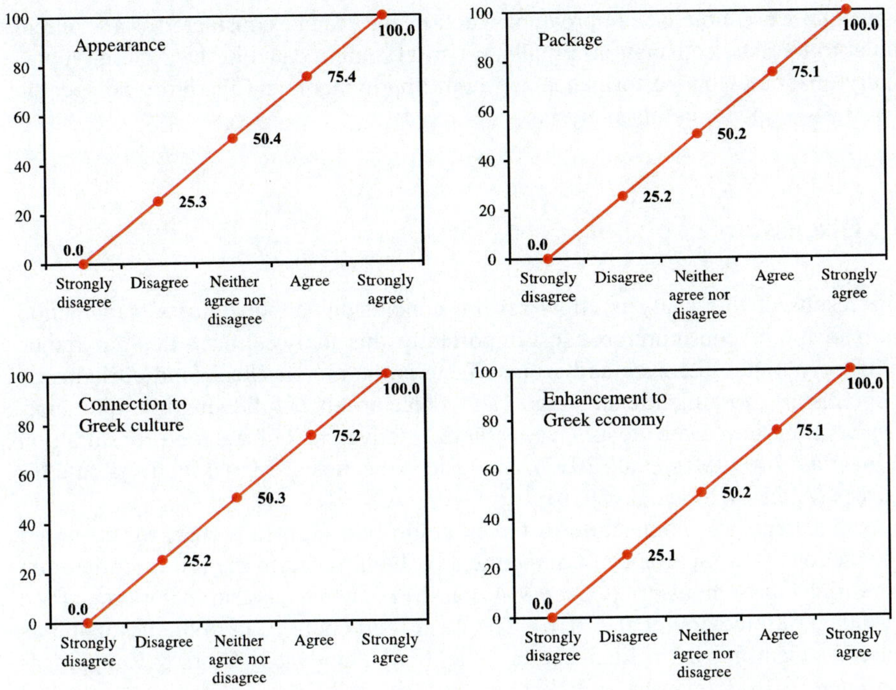

Fig. 2 (continued)

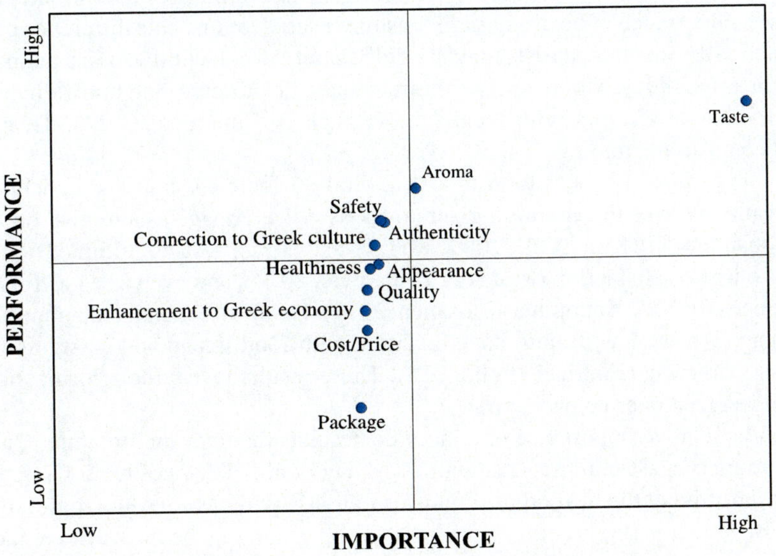

Fig. 3 Estimated action diagram

they require no immediate improvement action, they can be considered as a potential threat to tourists' satisfaction. Finally, no criteria are located in the Action Opportunity quadrant (low performance and high importance), and therefore no specific weaknesses appear in this analysis.

5 Discussion

The results of this analysis allow several conclusions about tourists' satisfaction from local food consumption. Most importantly, this study confirms the importance of sensory traits, like taste and aroma. Taste is the most critical food attribute, as it obtains the most significant weight, and a nice aroma is following in importance. Sensory traits have been identified as essential motivators for local food consumption (Kim et al., 2009; Mak et al., 2012), and taste is the most essential food attribute for tourists (Altintzoglou et al., 2016).

Authenticity and connection to Greek culture are also important for customer satisfaction. The importance of authenticity is highlighted in the relevant literature as a strong tourist motivator (Cohen & Avieli, 2004), while it is also an important food attribute (Altintzoglou et al., 2016). The local food connection to the destination's culture is identified in the literature as a vital tourist motivator (Björk & Kauppinen-Räisänen, 2014; Kim & Eves, 2012).

The importance of safety reflects the tourists' concerns that have been identified in the relevant literature (Cohen & Avieli, 2004). This finding seems contradictory to the lower importance of healthiness for customer satisfaction. This difference can be attributed to the fact that tourists may prioritize safety over healthiness in a temporary situation as a holiday. Moreover, in a similar study, healthiness, and nutrition did not affect tourist satisfaction with local food (Peštek & Činjarević, 2014). Thus, our results confirm this finding.

The appearance of local food is also a vital attribute for tourists' satisfaction. This result confirms the attention given to foods' visual image as a motive for local food consumption (Kim et al., 2009; Mak et al., 2012), as a food dimension that affects tourists' satisfaction (Peštek & Činjarević, 2014), and even as a gastronomic experience (Björk & Kauppinen-Räisänen, 2014). On the other hand, the importance of package is very low despite its association with food image and its significance for food marketing (Endrizzi et al., 2015). These results reveal the prioritization of food appearance over its packaging.

The low relative importance of quality contradicts the relevant literature. Quality affects tourists' gastronomic satisfaction (Akdag et al., 2018; Peštek & Činjarević, 2014), and it is a crucial food attribute for tourists' purchases (Altintzoglou et al., 2016). Beyond tourism literature, products quality is among the most critical factors for consumers who purchase local food (Stephenson & Lev, 2004). A possible explanation could be that quality is important for tourists, but it may be taken as granted, and this may results to a relatively low weight compared to other attributes.

The low weight of the enhancement of the local economy signals that sustainability concerns are not that important for tourists' satisfaction. Other studies have found that local communities' economic and social sustainability is important for tourists (Sims, 2009). Our study shows that the satisfactory effect of enhancing local communities is low, but this may be affected by the characteristics of the respondents includes in the sample.

Cost gets the lowest relative importance which is an interesting result but contradictory to the findings of the relevant literature. Price is a dimension of the gastronomic image of a destination; it can affect purchase intentions (Ahmad et al., 2019) and tourists' satisfaction (Peštek & Činjarević, 2014). One potential explanation could be that local food consumption is an experiential part of the holiday (Quan & Wang, 2004). Therefore, they are willing to pay a price premium to get new and fulfilling experiences (Morgan, 2006).

Concluding from the performance indices, tourists are delighted with local food consumption. At the same time, they mainly evaluate them as tasty, with a nice aroma, authentic, safe, with a nice appearance, and connected to Greek culture. They less consider them as inexpensive, having a nice package and enhancing the Greek economy. Concluding from the demanding indices, tourists are not demanding towards the selected criteria.

There are demographic differences in tourists' agreement with satisfaction with local food. Female respondents with a household size of two members, a monthly income of 3000 euro, and an average age of 39 years old stated that they are satisfied by consuming local food. Demographic traits affect tourists' decision to consume local food (Kim et al., 2009).

6 Conclusions

The presented study aims to measure the importance of the effect of local food attributes on tourists' satisfaction. One of the main advantages of the study is the application of an extension of the MUSA method in order to respect the qualitative nature of customer judgments, assuring robust results (Grigoroudis & Siskos, 2010). Our study contributes to the literature on tourists' satisfaction with local food (Chi & Qu, 2008, 2009; Peštek & Činjarević, 2014). Besides, it further adds to the relevant MUSA applications in tourism industry (see for example (Tsitsiloni et al., 2013; Delias et al., 2018)).

The results of the study may be used by agribusiness managers and food retailers in the tourism sector in order to sustain and improve tourists' satisfaction. Tourism stakeholders should emphasize taste, connection to Greek culture, safety, nice appearance, authenticity, and nice aroma, which are considered as major competitive advantages. On the other hand, healthiness, quality, enhancement to Greek economy, cost and nice package are classified as potential threats. Therefore, tourism stakeholders should try to increase the performance of these attributes, by communicating for example the health advantages of Greek food and their contribution to local development.

The main limitation of the study concerns the relatively small sample size. The population of tourists visiting Greece is rather heterogenous, and therefore future research may focus on specific tourist groups having distinctive geographic, demographic, socioeconomic, psychographic, or behavioral characteristics and preferences. Specifically, the MUSA method as a collective model provides aggregate measures that might mask latent heterogeneity. Future studies should aim at understanding differences in tourists' satisfaction and deriving market segments.

Another limitation of the study is related to the assessment of the satisfaction criteria. More specifically, tourists may have difficulties to understand the distinction between some criteria (e.g., differences between culture and authenticity or between aroma and taste), particularly regarding non-expert respondents. Therefore, potential interrelations among satisfaction criteria may appear. For this reason, future research may examine alternative extensions of the MUSA method that can consider possible criteria interactions (Angilella et al., 2014).

References

Ahmad, M. S., Jamil, A., Latif, K. F., Ramayah, T., Ai Leen, J. Y., Memon, M., & Ullah, R. (2019). Using food choice motives to model Pakistani ethnic food purchase intention among tourists. *British Food Journal, 122*(6), 1731–1753.

Akdag, G., Guler, O., Dalgic, A., Benli, S., & Cakici, A. C. (2018). Do tourists' gastronomic experiences differ within the same geographical region? A comparative study of two Mediterranean destinations: Turkey and Spain. *British Food Journal, 120*(1), 158–171.

Altintzoglou, T., Heide, M., & Borch, T. (2016). Food souvenirs: Buying behaviour of tourists in Norway. *British Food Journal, 118*(1), 119–131.

Angilella, S., Corrente, S., & Słowiński, R. (2014). MUSA-INT: Multicriteria customer satisfaction analysis with interacting criteria. *Omega, 42*(1), 189–200.

Björk, P., & Kauppinen-Räisänen, H. (2014). Culinary-gastronomic tourism: A search for local food experiences. *Nutrition & Food Science, 44*(4), 294–309.

Björk, P., & Kauppinen-Räisänen, H. (2016). Local food: A source for destination attraction. *International Journal of Contemporary Hospitality Management, 28*(1), 177–194.

Chang, R. C. Y., & Mak, A. H. N. (2018). Understanding gastronomic image from tourists' perspective: A repertory grid approach. *Tourism Management, 68*, 89–100.

Chi, C.G.-Q., & Qu, H. (2008). Examining the structural relationships of destination image, tourist satisfaction and destination loyalty: An integrated approach. *Tourism Management, 29*(4), 624–636.

Chi, C.G.-Q., & Qu, H. (2009). Examining the relationship between tourists' attribute satisfaction and overall satisfaction. *Journal of Hospitality Marketing & Management, 18*(1), 4–25.

Cohen, E., & Avieli, N. (2004). Food in tourism. *Annals of Tourism Research, 31*(4), 755–778.

Delias, P., Grigoroudis, E., & Matsatsinis, N. (2018). Developing regional strategies based on tourist behaviour analysis: A multiple criteria approach. In N. Matsatsinis & E. Grigoroudis (Eds.), *Preference disaggregation in multiple criteria decision analysis* (pp. 187–209). Springer.

Ellis, A., Park, E., Kim, S., & Yeoman, I. (2018). What is food tourism? *Tourism Management, 68*, 250–263.

Endrizzi, I., Torri, L., Corollaro, M. L., Demattè, M. L., Aprea, E., Charles, M., Biasioli, F., & Gasperi, F. (2015). A conjoint study on apple acceptability: Sensory characteristics and nutritional information. *Food Quality and Preference, 40*, 39–48.

Ghanem, M. (2019). The behavioral intention of tourists toward local foods: An applied research on the local foods served in Egyptian Siwa oasis. *Journal of Service Science and Management, 12*, 714–741.

Grigoroudis, E., & Politis, Y. (2015). Robust extensions of the MUSA method based on additional properties and preferences. *International Journal of Decision Support Systems, 1*(4), 438–460.

Grigoroudis, E., & Siskos, Y. (2002). Preference disaggregation for measuring and analysing customer satisfaction: The MUSA method. *European Journal of Operational Research, 143*(1), 148–170.

Grigoroudis, E., & Siskos, Y. (2010). *Customer satisfaction evaluation: Methods for measuring and implementing service quality*. Springer.

Grunert, K. G. (1997). What's in a steak? A cross-cultural study on the quality perception of beef. *Food Quality and Preference, 8*(3), 157–174.

Hammami, A. M., Stanton, J. L., Periklis, D., Baourakis, G., van Dijk, G., & Mamalis, S. (2018). Tourist's behaviour towards local Cretan food. In K. Mattas, G. Baourakis, & C. Zopounidis (Eds.), *Sustainable agriculture and food security* (pp. 49–61). Springer.

Henderson, J. C. (2009). Food tourism reviewed. *British Food Journal, 111*(4), 317–326.

Kim, Y. G., & Eves, A. (2012). Construction and validation of a scale to measure tourist motivation to consume local food. *Tourism Management, 33*(6), 1458–1467.

Kim, Y. G., Eves, A., & Scarles, C. (2009). Building a model of local food consumption on trips and holidays: A grounded theory approach. *International Journal of Hospitality Management, 28*(3), 423–431.

Kim, Y. H., Kim, M., & Goh, B. K. (2011). An examination of food tourist's behavior: Using the modified theory of reasoned action. *Tourism Management, 32*(5), 1159–1165.

Koutsimanis, G., Getter, K., Behe, B., Harte, J., & Almenar, E. (2012). Influences of packaging attributes on consumer purchase decisions for fresh produce. *Appetite, 59*(2), 270–280.

López-Guzmán, T., Uribe Lotero, C. P., Pérez Gálvez, J. C., & Ríos Rivera, I. (2017). Gastronomic festivals: Attitude, motivation and satisfaction of the tourist. *British Food Journal, 119*(2), 267–283.

Mak, A. H. N., Lumbers, M., Eves, A., & Chang, R. C. Y. (2012). Factors influencing tourist food consumption. *International Journal of Hospitality Management, 31*(3), 928–936.

Mak, A. H. N., Lumbers, M., Eves, A., & Chang, R. C. Y. (2017). The effects of food-related personality traits on tourist food consumption motivations. *Asia Pacific Journal of Tourism Research, 22*(1), 1–20.

Medina, F. (2009). Mediterranean diet, culture and heritage: Challenges for a new conception. *Public Health Nutrition, 12*(9A), 1618–1620.

Morgan, M. (2006). Making space for experiences. *Journal of Retail & Leisure Property, 5*(4), 305–313.

Perçin, N. S., Eren, D., & Sezgin, A. (2021). The effects of local food consumption motivations on customer satisfaction: A research on Turkish food. *Journal of Social Research and Behavioral Sciences, 7*(13), 400–423.

Peštek, A., & Činjarević, M. (2014). Tourist perceived image of local cuisine: The case of Bosnian food culture. *British Food Journal, 116*(11), 1821–1838.

Quan, S., & Wang, N. (2004). Towards a structural model of the tourist experience: An illustration from food experiences in tourism. *Tourism Management, 25*(3), 297–305.

Rehman, S. U., Samad, S., Singh, S., & Usman, M. (2022). Tourist's satisfaction with local food effect behavioral intention in COVID-19 pandemic: A moderated-mediated perspective. *British Food Journal, 124*(10), 3133–3151.

Sanchez-Cañizares, S., & Castillo-Canalejo, A. M. (2015). A comparative study of tourist attitudes towards culinary tourism in Spain and Slovenia. *British Food Journal, 117*(9), 2387–2411.

Sánchez-Cañizares, S. M., & López-Guzmán, T. (2012). Gastronomy as a tourism resource: Profile of the culinary tourist. *Current Issues in Tourism, 15*(3), 229–245.
Sims, R. (2009). Food, place and authenticity: Local food and the sustainable tourism experience. *Journal of Sustainable Tourism, 17*(3), 321–336.
Stephenson, G., & Lev, L. (2004). Common support for local agriculture in two contrasting Oregon communities. *Renewable Agriculture and Food Systems, 19*(04), 210–217.
Tikkanen, I. (2007). Maslow's hierarchy and food tourism in Finland: Five cases. *British Food Journal, 109*(9), 721–734.
Triantafillidou, A., Yannas, P., & Lappas, G. (2019). The impact of the destination image of Greece on tourists' behavioral intentions. In N. Sykianakis, P. Polychronidou, & A. Karasavvoglou (Eds.), *Economic and financial challenges for eastern Europe* (pp. 345–359). Springer.
Tsitsiloni, M., Grigoroudis, E., & Zopounidis, C. (2013). Service quality evaluation in the tourism industry: A SWOT analysis approach. Springer Proceedings in Mathematics & StatisticsIn A. Migdalas, A. Sifaleras, C. Georgiadis, J. Papathanasiou, & E. Stiakakis (Eds.), *Optimization theory, decision making, and operations research applications* (Vol. 31, pp. 249–266). Springer.
Wang, C., Qu, H., & Hsu, M. K. (2016). Toward an integrated model of tourist expectation formation and gender difference. *Tourism Management, 54*, 58–71.

Ecotourism as a Tool for Regional Development in the Area of Prespa National Forest Park

Christiana Koliouska, Zacharoula Andreopoulou, and Constantin Zopounidis

Abstract The promotion of the Prespa National Forest Park through ecotourism is very important for the local economic development of the region as it constitutes one of the most popular destinations for a high level of quality tourism throughout the year. Some factors that affect the promotion and the development of the tourism product are the following: the natural wealth, the natural landscape, the geographical part, the uniqueness of the local products, the possibility of creating new facilities and, in general, the bearing capacity of the region. The main element of ecotourism development is the benefit of small and medium enterprises, as well as tourists, from the new type of alternative tourism, preserving the authentic, ecological and natural beauty of the landscape. The aim of this paper is to study the area of the Prespes National Park as an ecotourism destination and to promote the provided alternative activities on the Internet. The websites that promote the tourism enterprises of the area are analyzed and classified using the method of multi-criteria analysis PROMETHEE II. Finally, suggestions for the optimization of the existing websites are presented in order to enhance the online promotion of the area of Prespa National Forest Park.

Keywords Ecotourism · Regional development · Prespa National Forest Park · Internet · Websites

C. Koliouska (✉) · Z. Andreopoulou
Laboratory of Forest Informatics, Faculty of Forestry and Natural Environment, Aristotle University of Thessaloniki, Thessaloniki, Greece
e-mail: ckolious@for.auth.gr

Z. Andreopoulou
e-mail: randreop@for.auth.gr

C. Zopounidis
Technical University of Crete, Crete, Greece

Audencia Business School, Nantes, France

C. Zopounidis
e-mail: kostas@dpem.tuc.gr

C. Zopounidis et al. (eds.), *Operational Research Methods in Business, Finance and Economics*, Lecture Notes in Operations Research,
https://doi.org/10.1007/978-3-031-31241-0_6

1 Introduction

The evolution of different forms of alternative tourism has significantly affected the economy and the demands of the environment. Since 1980, new forms of alternative tourism emerged as an alternative solution to replace mass tourism and serve the demands for differentiated and individualized tourism experiences. European Union (EU) always provides a particular set of support to the Member States regarding tourism development by funding projects to implement sustainable and local strategies (Chatzitheodoridis & Kontogeorgos, 2020).

EU strategy for alternative tourism promotion in Greece refers to domestic tourism, mainly in mountain areas and not in Greek islands. Mountain areas are characterized as areas with geographical disadvantages because they are not coastal areas, and so, they are not the major Greek tourism product. Some of the main reference points of an attractive tourism product are the following: natural resources, the landscape, the geographical area, the soil, the uniqueness of local products, the possibility of facility expansion and the carrying capacity of the area (Voudouris et al., 2018). In particular, Florina Regional Unit is considered one of the top ten agrotourism destinations in Greece and received an European Destinations of Excellence (EDEN) award from EU as a tourist destination that provides high-level quality of services and products. A recent research is referred to the importance of high-quality services provision as the motivation of the local enterprises seems to be the long-term stay and the persistence in visiting their destination (Christou & Chatzigeorgiou, 2020). Undoubtedly, these goals can be achieved only with the contribution of the residents (know-how and skills).

Recent studies indicate that the implemented strategies to increase the attractiveness of mountain areas in Greece seem to be ineffective. On the other side, winter tourism in Greece is enhanced by enriched web-based portal, funding marketing of special events or festivals and operating tourism promotion agencies (Smeral, 2006). Last years, there have been some efforts to introduce young people to the concepts of alternative forms of tourism and agrotourism through the Environmental Action Program for Schools Guide (EAPS). The main element of ecotourism product development is the advanced benefit of the innovative and diversified tourism product both for the enterprises and the tourist while maintaining a healthy environment, preserving the local culture and achieving a balance between local community and wildlife.

Tourism enterprises strive to be competitive and establish a strategic advantage, which is possessed through e-commerce (Andreopoulou et al., 2017a). The aim of each website is to provide rich information for a location and an interactive DB (Database) for the wider area in order to support and strengthen efforts for local development (Koliouska et al., 2021). The enterprises aim to be active and participate in the Information Society since the benefits are too many and the new Information and Communication Technologies (ICT) have the ability to serve customers without geographical restrictions, 24/7 (Tsekouropoulos et al., 2013), while at the same time, there is a cost reduction. Websites enable all enterprises to complete their commerce process and achieve more effective collaboration by acting as intermediaries between customers and suppliers (Saprikis & Vlachopoulou, 2012).

This paper studies the wider area of Prespa National Forest Park in Greece as an ecotourism destination and the provided alternative activities on the Internet. The websites that promote the tourism enterprises of the region are analyzed and classified using the PROMETHEE II multi-criteria analysis method. Some proposals are presented aiming to develop and optimize the websites that already exist in order to enhance the online promotion of the wider area of the Prespa National Forest Park.

2 Methodology

The search of the websites promoting ecotourism in the wider area of the Prespa National Forest Park in Greece was carried out in the Internet using the proper keywords and phrases in the large-scale search engine "Google" that gives very satisfactory results compared to any other existing search engine (Langville & Meyer, 2006). However, some alternative search engines were checked out to verify the results, such as "Yahoo", "Bing", "MSN Search", "Ask", and "Pathfinder".

The first step was to conduct a qualitative analysis in order to organize, code, explore and report on the collected data. By this way, the type of common characteristics identified in the websites can be examined. The second step was to conduct a quantitative analysis to determine the absence or presence of each characteristic on each website. Each characteristic represents a variable, xi. This paper studies 30 website characteristics, which are presented in Table 1 (Koliouska, 2013).

The PROMETHEE II method for multiple criteria decision-making was applied, which provides a final ranking of the alternatives. This method deals only with data with fixed numerical values and a reasonable degree of accuracy (Goumas & Lygerou, 2000). The ranking is based on the balance between the two preference functions. All alternatives are compared in pairs, based on the defined characteristics which constitute the evaluation criteria (Brans & Smet, 2016). The algorithm indicates the superiority between two alternatives according to the studied characteristics by applying six types of general criteria (Brans et al., 1986). The decision-maker has to conduct quantitative and qualitative assessments to determine the performance of the alternatives regarding the criteria (Li & Li, 2010). Moreover, the decision-maker has to determine the relative importance of the defined evaluation criteria that should reflect on the problem statement. Finally, the alternatives are ranked from the best to the worst according to their net flow score (φ)—the alternative with the higher score is supposed to be superior than the rest of the alternatives (Yilmaz & Dağdeviren, 2011). PROMEHEE II method is applied in all fields of science (Andreopoulou et al., 2017b; Pohekar & Ramachandran, 2003).

Table 1 Description of 30 website characteristics

Variable	Website characteristic
X1	Two or more languages
X2	Information about products services activities
X3	Contact information
X4	Local information
X5	Digital map
X6	Audiovisual material
X7	Live web camera
X8	Search engine
X9	Sitemap
X10	Updated enterprise information
X11	Online survey
X12	Online communication form
X13	Weather forecast
X14	Website visitor tracker
X15	Frequently asked questions (FAQ)
X16	Links to other companies, etc.
X17	Various topics of interest
X18	Downloadable files
X19	Calendar application
X20	Event calendar application
X21	Celebration calendar application
X22	Social media sharing
X23	Social media profile
X24	Forum
X25	Related sources of information
X26	Third person advertisement
X27	Newsletter
X28	RSS
X29	Code access
X30	Personalization of the page, trace, safety

3 Results

The research in the Greek Internet pointed out 18 websites that promote ecotourism in the wider area of Prespa National Forest Park. According to the results of PROMETHEE II methodology, net flow scores of ecotourism enterprise websites range from 0.79629380 to −0.985414989. There is a big difference in superiority

Table 2 Final ranking of ecotourism enterprises in Prespa National Forest Park using PROMETHEE II method

	Ecotourism enterprise	Net flow φ
1	EE_1	0.796293800
2	EE_2	0.757208910
3	EE_3	0.646713570
4	EE_4	0.638629573
5	EE_5	0.564122305
6	EE_6	0.558628917
7	EE_7	0.534854485
8	EE_8	−0.409183640
9	EE_9	−0.445341449
10	EE_10	−0.544099488
11	EE_11	−0.718855063
12	EE_12	−0.811107425
13	EE_13	−0.841188395
14	EE_14	−0.928956889
15	EE_15	−0.928956889
16	EE_16	−0.964153193
17	EE_17	−0.975414989
18	EE_18	−0.985414989

between the first and last ecotourism website in the final ranking. The website characteristics of the ecotourism enterprise with the highest net flow score are the following: the provision of the website content in more than two languages, the provision of information on the products/services, the provision of contact details, the provision of information of local interest, the provision of useful links, the provision of information on various topics, the provision of a digital map, the provision of photographic/audio-visual material, the maintenance of an updated website, the provision of an online contact form, the provision of a social network account (profile) and the provision of information on related issues and advertisements of third parties (Table 2).

4 Conclusion

This paper presents the new trends in the field of Internet promotion of the wider area of Prespa National Forest Park aiming to enhance ecotourism activities in the context of sustainable regional development. The research in the Greek Internet resulted in 18 websites of ecotourism enterprises in the wider area of Prespa National Forest Park. Most websites provide the possibility of viewing their content in Greek and English or only in Greek. Furthermore, the websites include information on the provided products/services and information on the enterprise (entity), activities and contact information, interactive content about the local area using images, videos, digital interactive maps, live web camera or 3D image, links to related topics and links to various topics, as well as information on related topics while also advertising third parties.

During the current smart era, the adoption of innovative technological changes is essential. The websites that promote the protected areas should evolve further and become more effective in order to enhance the provision of upgraded e-services, as well as the provision of reliable multimedia material. The integration of digital tools and services can lead to the transformation of a website into an effective and efficient means of advertising. Ecotourism websites that achieved a low net flow score can use as benchmarks the superior ecotourism websites, since they fulfill more characteristics and adopt more online tools and services that should be used at the website design (Andreopoulou et al., 2017b).

References

Andreopoulou, Z., Lemonakis, C., Koliouska, C., & Zopounidis, C. (2017a). Internet and agro-tourism sector for regional development in Crete: A multicriteria ranking. *International Journal of Information Decision Sciences, 9*(2), 116–127.

Andreopoulou, Z., Koliouska, C., & Zopounidis, C. (2017b). *Multicriteria and clustering: Classification techniques in agrifood and environment*. Springer.

Brans, J. P., & De Smet, Y. (2016). PROMETHEE methods. In *Multiple criteria decision analysis* (pp. 187–219). Springer.

Brans, J. P., Vinke, P., & Mareschal, B. (1986). How to select and how to rank projects: The PROMETHEE method. *European Journal of Operation Research, 24*, 228–238.

Chatzitheodoridis, F., & Kontogeorgos, A. (2020). Exploring of a small-scale tourism product under economic instability: The case of a Greek rural border area. *Economies, 8*(3), 52.

Christou, E., & Chatzigeorgiou, C. (2020). Adoption of social media as distribution channels in tourism marketing: A qualitative analysis of consumers' experiences. *Journal of Tourism, Heritage & Services Marketing (JTHSM), 6*(1), 25–32.

Goumas, M., & Lygerou, V. (2000). An extension of the PROMETHEE method for decision making in fuzzy environment: Ranking of alternative energy exploitation projects. *European Journal of Operational Research, 123*(3), 606–613.

Koliouska, C. (2013). Assessment of internet marketing and promotion of the National Parks in Greece. Master dissertation Department of Forestry and Natural Environment, Aristotle University of Thessaloniki.

Koliouska, C., Andreopoulou, Z., Doumpos, M., Galariotis, E., & Zopounidis, C. (2021). Multicriteria evaluation of the websites of alternative tourism enterprises: Case study in the region of Crete. *IEEE Transactions on Engineering Management*.

Langville, A. N., & Meyer, C. D. (2006). *Google's pagerank and beyond: The science of search engine rankings*. Princeton University Press

Li, W. X., & Li, B. Y. (2010). An extension of the Promethee II method based on generalized fuzzy numbers. *Expert Systems with Applications, 37*(7), 5314–5319.

Pohekar, S. D., & Ramachandran, M. (2003). Application of multi-criteria decision making to sustainable energy planning—A review. *Renewable and Sustainable Energy Reviews, 8*.

Saprikis, V., & Vlachopoulou, M. (2012). Determinants of suppliers' level of use of B2B E-marketplaces. *Industrial Management & Data Systems, 112*(4), 619–643.

Smeral, E. (2006). Aspects to justify public tourism promotion: An economic perspective. *Tourism Review*.

Tsekouropoulos, G., Andreopoulou, Z., Koliouska, C., Koutroumanidis, T., Batzios, C., & Samathrakis, V. (2013). Internet functions in marketing: Multicriteria ranking of agricultural SMEs websites in Greece. *Journal of Agricultural Informatics, 4*(2), 22–36.

Voudouris, K., Mandrali, P., & Kazakis, N. (2018). Preventing groundwater pollution using vulnerability and risk mapping: The case of the Florina Basin, NW Greece. *Geosciences, 8*(4), 129.

Yilmaz, B., & Dağdeviren, M. (2011). A combined approach for equipment selection: F-PROMETHEE method and zero–one goal programming. *Expert Systems with Applications, 38*(9), 11641–11650.

Evaluating the Importance of ESG Criteria: A Multicriteria Approach

Marianna Eskantar, Amal Aouadi, Karima Bouaiss, Constantin Zopounidis, and Michalis Doumpos

Abstract ESG criteria are considered an important topic that all businesses worldwide should consider. Businesses should be able to publish this kind of information so that you can consider its viability in matters concerning the three pillars. The big question that arises is whether all companies regardless of the sector or country they belong to should publish the same ESG indicators. In this work, the ESG criteria were examined over time, dividing the period 2007–2019 into three equal periods, for each pillar separately, for 39 countries worldwide (developed and developing) and 16 industries.

Keywords Sustainable finance · Environmental performance · Social impact · Corporate governance · Multicriteria analysis

1 Introduction

Climate change, social divides, democratic institutions, and the control of corporate management policies, are considered serious threats in recent decades to the sustainability of the planet and the well-being of societies (Yang et al., 2022). For this reason, considerable attention has been paid to these problems by researchers and regulators (Razzaq et al., 2021). In September 2015, the United Nations enacted the plan entitled "Transforming Our Word: The 2030 Agenda for Sustainable Development", considering the five P's: planet, people, prosperity, peace, and partnership (Khaled et al., 2021).

M. Eskantar (✉) · C. Zopounidis · M. Doumpos
School of Production Engineering & Management, Financial Engineering Laboratory, Technical University of Crete, University Campus, 73100 Chania, Greece
e-mail: meskantar@isc.tuc.gr

A. Aouadi
Audencia Business School, Nantes, France

K. Bouaiss · C. Zopounidis
IAE Lille, University School of Management, Lille, France

C. Zopounidis et al. (eds.), *Operational Research Methods in Business, Finance and Economics*, Lecture Notes in Operations Research,
https://doi.org/10.1007/978-3-031-31241-0_7

It all started with the concept of social responsibility, which is a much-used term that became popular during the 1980s and 1990s, but its roots can be traced back two millennia, shaped by civil rights era thinkers, and organizations that they rely on, faith and women. The modern process of social responsibility is based on three pillars:

1. Social issues to avoid based on values.
2. Sustainability-focused elements—commonly referred to as "Environmental Issues."
3. Corporate engagement and impact investing.

By the mid-2000s in Europe, three main catalysts created the demand for ESG analysis by large investors. The first was a vigorous intellectual and legal debate about the relationship between fiduciary duty and sustainability issues. The second was climate change. The third was a capitulation to the position that bad corporate governance was harmful to markets.

The growth of ESG (environmental, social and governance) investments has stimulated general interest among asset managers. In 2019, the capitalization of ESG-focused portfolios in major markets exceeded US$30 trillion. Investors are interested in investing in ESG for at least two reasons. First, by focusing on ESG investing, ethical investment practices are actively promoted. Second, when ESG investments increase, the performance of a managed portfolio is assumed to improve, as returns increase and portfolio risk decreases.

The performance of a company is no longer assessed based solely on financial data, but also considering its contribution to environmental sustainability, the well-being of the surrounding community, and the corporate policy it implements (Zhao et al., 2018).

Disclosure of ESG information refers to the disclosure of a company's environmental, social, intergovernmental, and financial information in a comprehensive, timely and accurate manner so that the market can make a rational judgment on the value of the investment based on the interests of shareholders, creditors, and investors (Zhao et al., 2018). Since 1992, the Financial Initiative of the United Nations Environment Program has advocated for financial institutions to integrate ESG factors into their decision-making process.[1] Since then, ESG has gradually become one of the three main dimensions for measuring corporate performance and risk.

Since then, companies have been paying close attention to their social and environmental responsibilities as well as the governance policies they follow. The trust and support gained in the implementation of ESG practices, are likely to have a positive impact on the success and growth of companies (Aupperle et al., 1985; Giese et al., 2019).

Each ESG pillar is evaluated through various indicators. More specifically, commonly used indicators related to the environmental performance of a company include carbon emissions, greenhouse gas emissions, the use of renewable energy, and waste disposal, among others. Social performance is evaluated by criteria such

[1] https://bit.ly/3zdNp4d.

as child labor, discrimination, diversity (e.g., employee/board diversity), and other. Finally, corporate governance is measured using criteria such as board selection for executive compensation, executive compensation, the relationship between corporate stakeholders, power of directors, etc. (Escrig-Olmedo et al., 2019; Muñoz-Torres et al., 2019; Syed, 2017).

Naturally, many questions arise. For instance:

- What are the appropriate ESG criteria that a company must implement and publish to be considered sustainable?
- Is there a common standard for applying ESG criteria?
- Has the importance of the ESG pillars changed over the years?
- Should companies from developing countries manage the same ESG issues as companies from developed economies?

All these questions and many more are slowly being answered by the scientific community for better management of sustainability issues when the moment of implementation and decision-making comes. The objective of this study is to provide empirical results on these issues through the application of a multicriteria decision making methodology, based on a large cross-country sample.

The chapter is organized as follows. Section 2 presents the multi-criteria methodology used in the analysis. Section 3 describes the data and indicators, whereas Sect. 4 presents the results. Finally, Sect. 5 concludes the chapter and discusses some future research directions.

2 Methodology

The analysis is based on a multicriteria decision aiding (MCDA) approach, based on the principles of preference disaggregation analysis (Jacquet-Lagrèze & Siskos, 2001). MCDA deals with procedures and methodologies for supporting the decision-making process in problems with multiple (conflicting) criteria. The PDA paradigm in MCDA facilitates the development of decision support models, following a regression-based approach. PDA methods enable the analysis of global preference judgements to identify the underlying decision model that best fits a set of input data.

In the present study, we employ a variant of the UTA method (Jacquet-Lagreze & Siskos, 1982). In the context of the UTA method, the preference judgements that are used to construct a decision model, are given in the form of an ordinal ranking. In this study, however, ESG scores are employed, which are continuous numerical variables, usually expressed in a 0–100 scale.

The evaluation model considered in the proposed setting is expressed in the form of an additive value function:

$$V(\mathbf{x}_i) = \sum_{j=1}^{n} w_j v_j\left(x_{ij}\right) \tag{1}$$

where $\mathbf{x}_i = (x_{i1}, x_{i2}, \ldots, x_{in})$ is the data vector for observation i (e.g., firm) over n attributes (e.g., ESG indicators), w_j is the trade-off constant (i.e., weight) of attribute j, and $v_j(x_{ij})$ is the marginal value of observation i on attribute j. In the additive model (1), $V(\mathbf{x}_i)$ represents an overall performance for each observation i, measured on a scale from 0 to 100, with higher values indicating higher ESG performance. The marginal value functions $v_j(\cdot)$, $j = 1, \ldots, n$, provide a decomposition of the overall assessment into partial evaluations for each individual attribute. These partial assessments are defined on a 0–100 scale, with higher values representing higher ESG performance on the corresponding attributes. In accordance with standard MCDA principles, the marginal value functions are monotone with respect to the attributes' levels, i.e., non-decreasing for attributes that increase ESG performance and non-increasing for attributes that have a negative impact on ESG. Moreover, the attributes' weights are defined to be non-negative and normalized such that they sum up to 1.

In accordance with Doumpos et al. (2017) the additive model (1) can be extended to include nominal effects, which relate to categorical attributes that do not have a clear monotonic relationship with the global preference evaluations under consideration (i.e., ESG scores):

$$V(\mathbf{x}_i, \mathbf{f}_i) = \sum_{j=1}^{n} w_j v_j(x_{ij}) + \mathbf{b}^\top \mathbf{f}_i \tag{2}$$

where the vector $\mathbf{f}_i$ consists of the data on the categorical variables for observation i and $\mathbf{b}$ is the vector of the corresponding coefficients, which are unrestricted in sign.

Assuming a reference (training) set of m observations for firms, described over n ESG indicators and having global ESG scores $y_1, y_2, \ldots, y_m$, the construction of the additive model (2) is performed through the solution of the following optimization problem:

$$\begin{aligned}
\min \quad & \lambda \sum_{i=1}^{m} (\sigma_i^+ + \sigma_i^-) + \left(||\mathbf{w}||^2 + ||\mathbf{b}||^2\right) \\
\text{s.t.}: \quad & \sum_{j=1}^{n} w_j v_j(x_{ij}) + \mathbf{b}^\top \mathbf{f}_i + \sigma_i^+ - \sigma_i^- = y_i \quad i = 1, 2, , \ldots, m \\
& v_j(\cdot) \text{monotonic} \qquad j = 1, \ldots, n \\
& w_j, \sigma_i^+, \sigma_i^- \geq 0, \mathbf{b} \in \mathbb{R}
\end{aligned} \tag{3}$$

where σ_i^+ and σ_i^- are error variables representing the deviations between the estimated and the actual ESG score y_i for observation i, and $\lambda > 0$ is a user-defined constant. The objective combines the fitting error with a regularization term for the parameters of the model. The weighting parameter λ defines the trade-off between these two objectives.

It should be noted that in the solution of the optimization model (3), the optimal weights $w_1, \ldots, w_n$ are not normalized to sum up to 1; the normalization is done by dividing the optimal weights by their sum. Moreover, the constraints in the above formulation appear in nonlinear form (as both the attributes' weights and the marginal value functions are unknown), but they can be expressed in linear form, assuming that the marginal value functions are piecewise linear, as in the UTA family of methods (for details, see Doumpos & Zopounidis, 2002; Siskos et al., 2005).

A model similar to (3) with purely ordinal data, was also considered by Angelopoulos et al. (2019) in a context of time series forecasting in energy planning. However, model (3) enables the consideration of categorical variables, whereas the addition of the regularization term in the objective improves the robustness of the obtained results in accordance with the principles of ridge regression and Tikhonov regularization (Tikhonov et al., 1995).

3 Data

The data concern ESG criteria (Tables 1, 2 and 3) which were collected based on the literature and the availability of the data. The sample comes from Thomson Reuters and covers the period 2007–2019, 16 sectors (Table 1), and 39 countries (Table 2). Overall, the sample consists of 17,017 firm-year observations in an unbalanced panel.

The environmental criteria are refer to two categories as shown in Table 3. In more detail:

- E1 describes if the company claims to have an ISO 14000 or EMS certification.

Table 1 Sample composition by NACE sector

Sector	Observations
A—Agriculture, forestry, and fishing	177
B—Mining and quarrying	1379
C—Manufacturing	6871
D—Electricity, gas, steam, and air conditioning supply	1035
E—Water supply; sewerage; waste management; remediation activities	150
F—Construction	393
G—Wholesale and retail trade	809
H—Transporting and storage	901
I—Accommodation and food service activities	193
J—Information and communication	1254
K—Financial and insurance activities	2338
L—Real estate activities	954
M—Professional, scientific, and technical activities	161
N—Administrative and support service activities	219
Q—Human health and social work activities	91
R—Arts, entertainment, and recreation	92

Table 2 Sample composition by country (no. of firm-year observations)

Country	Count	Country	Count	Country	Count
Argentina	41	Mexico	127	Denmark	204
Australia	711	Netherlands	258	Finland	260
Austria	127	New Zealand	55	France	872
Belgium	156	Norway	216	Germany	642
Brazil	196	Philippines	101	Greece	87
Canada	783	Poland	129	Hong Kong	576
Chile	69	Portugal	72	India	405
China	287	Russia	253	South Africa	559
Colombia	71	Singapore	172	South Korea	543
Indonesia	139	Taiwan	493	Spain	405
Italy	360	Thailand	221	Sweden	389
Japan	1507	Turkey	178	Switzerland	450
Malaysia	195	UK	1380	US	3328

Table 3 Environmental criteria

Category	Criteria
Emissions	E1: ISO 14000 or EMS
Emissions	E2: CO_2 equivalent emissions total
Emissions	E3: Biodiversity impact reduction
Resource use	E4: Land use
Resource use	E5: Policy environmental supply chain
Resource use	E6: Toxic chemicals reduction
Resource use	E7: Policy sustainable packaging
Resource use	E8: Total energy use to revenues ($ in million)

- E2 refers to the total amount of carbon dioxide (CO_2) and CO_2 equivalent emissions in tons.
- E3 indicates whether a company reports on its impact on biodiversity or on activities to reduce its impact on the native ecosystems and species, as well as the biodiversity of protected and sensitive areas.
- E4 indicates whether a company takes initiatives to reduce environmental impacts on land it owns, leases, or manages for productive activities or extractive use.
- E5 describes whether the company has a policy to include its supply chain in the company's efforts to reduce its overall environmental impact.
- E6 refers to whether a company reports initiatives to reduce, reuse, substitute or phase out toxic chemicals or substances.
- E7 refers to whether a company has a policy to improve the use of sustainable packaging.

Table 4 Social criteria

Category	Criteria
Workforce	S1: Policy Employee Health & Safety
Workforce	S2: Policy Supply Chain Health & Safety
Workforce	S3: Supply Chain Health & Safety Improvements
Workforce	S4: Training and Development Policy
Workforce	S5: Policy Diversity and Opportunity
Human Rights	S6: Policy Human Rights
Product responsibility	S7: Policy Data Privacy
Product responsibility	S8: Policy Responsible Marketing
Product responsibility	S9: Healthy Food or Products
Product responsibility	S10: ISO 9000
Product responsibility	S11: Product Responsibility Monitoring

- E8 measures the total direct and indirect energy consumption (in gigajoules) as a ratio of net revenue (in million US dollars).

The social criteria are presented in Table 4. All criteria in this pillar are expressed in a 0/1 binary format indicating whether a company has policies related to:

- S1: improve employee health and safety.
- S2: improve employee health and safety in its supply chain.
- S3: monitor the improvement of employee health and safety in its supply chain through surveys or measurements.
- S4: support the skills training or career development of employees.
- S5: promote diversity and equal opportunity.
- S6: ensure the respect of human rights.
- S7: protect customer and public privacy and integrity.
- S8: responsible marketing ensuring protection of children.
- S9: develop or market products and services that foster specific health and safety benefits for the consumers (healthy, organic, or nutritional food, safe cars, etc.).
- S10: existence of ISO 9000 certification or any industry specific certification (QS-9000-automotive, TL 9000-telecommunications, AS9100-aerospace, ISO/TS 16949-automotive, etc.).
- S11: monitoring the impact of products or services on consumers and the community.

The third pillar of ESG is governance. From this pillar we examined eight criteria (Table 5), which cover the following issues:

Table 5 Governance criteria

Category	Criteria
Management	G1: Policy board independence
Management	G2: Audit board committee
Management	G3: Policy board experience
Management	G4: Audit committee nonexecutive members
Management	G5: Audit committee independence
Shareholders	G6: Policy shareholder engagement
Shareholders	G7: Director election majority requirement
CSR strategy	G8: CSR sustainability external audit

- G1 examines whether the company has a policy regarding the independence of its board of directors.
- G2 checks for whether the company has an audit committee.
- G3 criterion draws conclusions about whether the company has a policy regarding sufficient experience on its board of directors.
- G4 examines the percentage of non-executive board members in the audit committee as defined by the company.
- G5 refers to the percentage of independent board members in the audit committee as defined by the company.
- G6 answers the question of whether the company has a policy to facilitate shareholder participation, resolutions, or proposals.
- G7 presents results that reveal whether company board members are generally elected by a majority.
- G8 checks whether the company has an external auditor of the CSR/H&A/Sustainability Report.

4 Results

This section presents and discusses the estimation results obtained through the proposed multicriteria methodology for the weights of the ESG criteria. The multi-criteria approach presented in Sect. 2 was implemented with different data specifications and the results are presented accordingly in this section. More specifically, in a first step, different estimations were obtained for each business sector in the sample, using the ESG scores in the Thomson Reuters' database as the dependent variable, the aforementioned indicators as the ESG performance criteria, as well as the years and countries as nominal variables. In the second level of the analysis, the previous analysis by sector was repeated separately for developing and developed countries, to examine the existence of possible differences between the two groups. Finally, the analysis by sector was performed over different time periods to examine the changes over time, using the countries as the only nominal variable. It should be noted that due to the small number of observations for some sectors for the partitions in the second and third steps of the analysis (i.e., classification of the countries by their level of economic development, and partition of the sample by time-period), the sectors have been clustered into larger groups with enough cases in each one to allow the derivation of meaningful results. The grouping of the sectors was performed based on their similarity in terms of each sector's activities.

4.1 Results by Sector

The first analysis of the results focuses on the differences between the 16 sectors presented in Table 1, for whole period under consideration (i.e., 2007–2019). In this step of the analysis, the multicriteria methodology was applied separately for each sector and the obtain results are summarized in this sub-section.

Table 6 shows the results of the environment pillar for each sector. The last column of the table presents average weights across all sectors. On average, the weight of the environmental dimension is 0.327, being most significant for sectors D (electricity, gas, steam, and air conditioning supply) and F (construction). According to the results, for agriculture, forestry, and fishing, the most important criterion is "Toxic Chemicals Reduction" (E6) with a weight of 0.113, while "Total Energy Use" (E8) is the least important criterion. For sector E (Water supply, sewerage, waste management, remediation activities) the criterion with a highest weight is "ISO 14000 or EMS" (E1), whereas for the construction sector (sector F) "CO_2 Equivalent Emissions "(E2) and "Policy Environmental Supply Chain" (E5) are the two most important indicators. "CO_2 Equivalent Emissions" (E2) is also the most important criterion for sector J (information and communication), whereas "Total Energy Use" (E8) is the most important criterion for sector Q (Human health and social work activities). On average, the most important criteria involve the total "CO_2 Equivalent Emissions" (E2), the "ISO 14000 or EMS" (E1), and "Policy Environmental Supply Chain" (E5).

The results for the social pillar are shown in Table 7. Regarding sector E (water supply, sewerage, waste management, remediation activities), the criterion with the highest weight is "Policy Human Rights" (S6) (0.112). For construction (sector F), the criterion "Supply Chain Health & Safety Improvements" (S3) (0.124) has the highest weight, whereas for sector industry J (Information and communication), "Training and Development Policy" (S4) is the most important factor. The same applies to sector R (arts, entertainment, and recreation). In conclusion, in this pillar we observe that the criteria "Training and Development Policy" (S4) and "Policy Human Rights" (S6) are the most important.

The results of the governance pillar are presented in Table 8. For wholesale and retail trade (sector G), "Audit Committee Non-Executive Members" (G4) and "CSR Sustainability External Audit" (G8) have the highest weight. Regarding sector K (financial and insurance activities), "Audit Board Committee" (G2) and "Audit Committee Independence" (G5) are the most important indicators, whereas for sector R (arts, entertainment, and recreation), important factors are "Audit Committee Independence" (G5) and "Policy Board Independence" (G1). It is worth noting that, on

Table 6 Weights of the criteria in the environmental pillar by sector

	A	B	C	D	E	F	G	H	I	J	K	L	M	N	Q	R	Avg
E1	0.076	0.033	0.063	0.076	0.128	0.000	0.009	0.031	0.043	0.022	0.060	0.018	0.071	0.082	0.077	0.055	0.053
E2	0.059	0.026	0.046	0.084	0.032	0.112	0.058	0.000	0.011	0.167	0.010	0.093	0.078	0.069	0.008	0.085	0.059
E3	0.014	0.050	0.041	0.075	0.046	0.038	0.025	0.047	0.059	0.027	0.005	0.060	0.081	0.004	0.000	0.016	0.037
E4	0.047	0.034	0.025	0.034	0.000	0.055	0.000	0.000	0.000	0.009	0.000	0.000	0.000	0.000	0.000	0.056	0.016
E5	0.059	0.048	0.056	0.062	0.049	0.107	0.064	0.032	0.088	0.049	0.076	0.076	0.071	0.085	0.083	0.017	0.064
E6	0.113	0.000	0.025	0.046	0.030	0.042	0.016	0.044	0.066	0.011	0.029	0.000	0.017	0.021	0.070	0.067	0.037
E7	0.028	0.000	0.027	0.002	0.000	0.021	0.061	0.000	0.003	0.032	0.000	0.039	0.000	0.000	0.074	0.011	0.019
E8	0.000	0.024	0.009	0.044	0.045	0.061	0.000	0.093	0.000	0.000	0.038	0.073	0.089	0.085	0.103	0.024	0.043
Total	0.396	0.214	0.291	0.423	0.330	0.437	0.232	0.247	0.270	0.317	0.220	0.358	0.407	0.347	0.415	0.330	0.327

Table 7 Weights of the criteria in the social pillar by sector

	A	B	C	D	E	F	G	H	I	J	K	L	M	N	Q	R	Avg
S1	0.069	0.023	0.025	0.070	0.069	0.081	0.025	0.048	0.018	0.033	0.042	0.030	0.093	0.007	0.051	0.078	0.048
S2	0.047	0.022	0.034	0.006	0.000	0.027	0.032	0.048	0.050	0.027	0.050	0.032	0.052	0.000	0.004	0.080	0.032
S3	0.000	0.054	0.000	0.000	0.000	0.124	0.024	0.007	0.001	0.000	0.000	0.000	0.000	0.081	0.000	0.000	0.018
S4	0.062	0.087	0.072	0.038	0.012	0.062	0.104	0.070	0.097	0.108	0.059	0.058	0.031	0.032	0.000	0.126	0.064
S5	0.034	0.040	0.043	0.055	0.005	0.012	0.040	0.043	0.070	0.043	0.048	0.020	0.000	0.049	0.000	0.035	0.034
S6	0.054	0.082	0.065	0.047	0.112	0.032	0.045	0.054	0.031	0.082	0.054	0.055	0.073	0.060	0.036	0.001	0.055
S7	0.031	0.030	0.030	0.004	0.015	0.028	0.055	0.055	0.050	0.023	0.077	0.053	0.006	0.000	0.086	0.030	0.036
S8	0.000	0.065	0.031	0.019	0.000	0.000	0.018	0.000	0.000	0.046	0.000	0.038	0.000	0.071	0.000	0.020	0.019
S9	0.000	0.000	0.020	0.030	0.000	0.035	0.046	0.022	0.029	0.000	0.000	0.027	0.000	0.000	0.054	0.000	0.016
S10	0.071	0.035	0.019	0.033	0.030	0.010	0.000	0.036	0.069	0.021	0.046	0.015	0.044	0.072	0.034	0.000	0.033
S11	0.040	0.041	0.050	0.027	0.074	0.000	0.047	0.072	0.032	0.045	0.021	0.005	0.073	0.038	0.076	0.047	0.043
Total	0.407	0.479	0.390	0.329	0.317	0.410	0.436	0.456	0.446	0.428	0.397	0.334	0.372	0.411	0.341	0.417	0.398

average, the governance pillar appears to be the least important one among the three ESG dimensions, with an average weight of 0.275 versus 0.398 for the social pillar and 0.327 for the environmental one.

Regarding the fit of the models, Table 9 presents the results of the coefficient of determination (R^2). It is evident that in all cases, R^2 is quite high, ranging from 80.7% (sector J) to 95.8% (sector R).

4.2 Results by Country Group

A second level of analysis involves the application of the multicriteria methodology to the data for two major groups of countries, defined based on their level of economic development. More specifically, the countries are categorized as developing or developed, using the classification by the United Nations' World Economic Situation and Prospects.[2]

Table 10 shows that there are no striking differences between developed and developing countries. Regarding the environmental dimension, total energy use is the most important criterion for both groups of countries, although with different weights in each case (developed: 0.406, developing: 0.298) as well as the least important pillar criterion which is "CO_2 Equivalent Emissions" (E2) (zero weight). The importance of the social criteria is very similar for the two groups of countries. The training and development policy is the most important factor in this pillar. Finally, regarding governance, the weights on this pillar for the two groups of countries show differences regarding the most important criterion. For the developed countries the "Audit Committee Independence" criterion is the most important factor, while in developing countries "Audit Committee Non-Executive Members" (G4) appears to be the most significant criterion.

Figure 1 presents the averages of the ESG pillars for developed and developing countries. For developed countries, more emphasis is placed on the social pillar, followed by the governance and environmental pillars. In contrast, for developing countries, the most important pillar is the environment, then governance, and finally the social pillar.

4.3 Results Over Time

The last part of the analysis focuses on the examination of the changes in the importance of the ESG indicators over different time periods. To this end, the multicriteria methodology is applied on three different time windows, each covering a period of five years, namely 2007–2011, 2011–2015, and 2015–2019. The discussion in this

[2] https://bit.ly/2DaSmxn.

Table 8 Weights of the governance criteria by sector

	A	B	C	D	E	F	G	H	I	J	K	L	M	N	Q	R	Avg
E1	0.076	0.033	0.063	0.076	0.128	0.000	0.009	0.031	0.043	0.022	0.060	0.018	0.071	0.082	0.077	0.055	0.053
E2	0.059	0.026	0.046	0.084	0.032	0.112	0.058	0.000	0.011	0.167	0.010	0.093	0.078	0.069	0.008	0.085	0.059
E3	0.014	0.050	0.041	0.075	0.046	0.038	0.025	0.047	0.059	0.027	0.005	0.060	0.081	0.004	0.000	0.016	0.037
E4	0.047	0.034	0.025	0.034	0.000	0.055	0.000	0.000	0.000	0.009	0.000	0.000	0.000	0.000	0.000	0.056	0.016
E5	0.059	0.048	0.056	0.062	0.049	0.107	0.064	0.032	0.088	0.049	0.076	0.076	0.071	0.085	0.083	0.017	0.064
E6	0.113	0.000	0.025	0.046	0.030	0.042	0.016	0.044	0.066	0.011	0.029	0.000	0.017	0.021	0.070	0.067	0.037
E7	0.028	0.000	0.027	0.002	0.000	0.021	0.061	0.000	0.003	0.032	0.000	0.039	0.000	0.000	0.074	0.011	0.019
G8	0.066	0.052	0.052	0.066	0.001	0.054	0.082	0.056	0.024	0.035	0.056	0.080	0.000	0.030	0.036	0.017	0.044
Total	0.396	0.214	0.291	0.423	0.330	0.437	0.232	0.247	0.270	0.317	0.220	0.358	0.407	0.347	0.415	0.330	0.327

Table 9 Coefficient of determination R^2 by sector

A	B	C	D	E	F	G	H	I	J	K	L	M	N	Q	R
0.948	0.844	0.819	0.875	0.907	0.873	0.879	0.843	0.927	0.807	0.819	0.832	0.916	0.914	0.936	0.958

Table 10 The weights of the ESG criteria for developed and developing countries

Environmental pillar			Social pillar			Governance pillar		
	Developed	Developing		Developed	Developing		Developed	Developing
E1	0.043	0.050	S1	0.037	0.042	G1	0.019	0.018
E2	0.006	0.043	S2	0.037	0.024	G2	0.025	0.018
E3	0.039	0.025	S3	0.012	0.006	G3	0.023	0.020
E4	0.008	0.027	S4	0.076	0.050	G4	0.070	0.055
E5	0.062	0.059	S5	0.035	0.047	G5	0.072	0.067
E6	0.026	0.028	S6	0.061	0.044	G6	0.030	0.020
E7	0.023	0.037	S7	0.043	0.039	G7	0.039	0.038
E8	0.047	0.054	S8	0.018	0.021	G8	0.051	0.071
			S9	0.018	0.037			
			S10	0.037	0.025			
			S11	0.045	0.036			

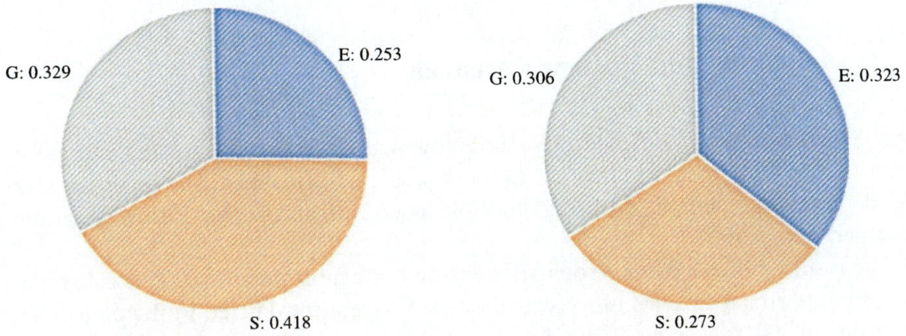

(a) Developed countries (b) Developing countries

Fig. 1 The average weights for the ESG for developed and developing countries

sub-section focuses on the relative importance of the three ESG pillars by business sector over these three time periods.

The obtained results are summarized in Table 11, for the clusters of sectors used in the analysis.[3] Starting from the environment pillar, its overall importance has decreased in the periods after 2011. This decrease is more evident in sectors A–B (agriculture and mining), J (information and communication), and K (financials). However, the importance of the environmental pillar has steadily increased for the manufacturing and construction sectors (C, F). The social pillar is the most important one in all periods, with small changes over time. The relative importance of

[3] As noted earlier, due to the small number of observations for some sectors in the separate time windows, the sectors have been clustered into larger groups with enough cases in each one.

Table 11 Criteria weights per pillar over the three time periods of the analysis

	2007–2011			2011–2015			2015–2019		
Sectors	E	S	G	E	S	G	E	S	G
A, B	0.375	0.374	0.251	0.274	0.447	0.279	0.219	0.468	0.313
C, F	0.230	0.409	0.360	0.262	0.430	0.309	0.311	0.377	0.312
D, E	0.409	0.311	0.280	0.323	0.377	0.300	0.392	0.362	0.246
G, H	0.220	0.430	0.351	0.176	0.487	0.337	0.205	0.448	0.347
I, L, M, N, Q, R	0.276	0.321	0.402	0.465	0.323	0.213	0.354	0.269	0.377
J	0.285	0.432	0.283	0.245	0.421	0.335	0.225	0.456	0.320
K	0.323	0.398	0.279	0.142	0.447	0.411	0.210	0.386	0.404
Mean	0.303	0.382	0.315	0.269	0.419	0.312	0.274	0.395	0.331

the governance pillar also shows minor variations over time. Nevertheless, a small increase in its total weight is evident in the most recent period.

5 Conclusions and Future Research

This is the first time a multicriteria methodology has been used to explain the importance of ESG pillars and indicators in a comprehensive manner. The analysis was based on a large sample covering multiple countries and sectors, as well as an extended time period.

The results indicate that the social dimension is the most important. This is logical as corporate social responsibility was the first concept introduced in the area. The environment pillar was also found to be of high importance, thus indicating that climate change is now affecting corporate decisions and investors. As far as the governance pillar is concerned, it appears to be of lower importance compared to the other two pillars.

Striking differences between developed and developing where not observed. Nevertheless, it is worth emphasizing the fact that the biggest differences between the two groups of countries are in the pillar of governance. Of course, the result seems to be reasonable, as in developing countries the issue of corruption and unethical behavior is of major interest.

Regarding the examination of the trends over time, we noticed that the importance of the environmental pillar has not changed much. Theoretically, this result seems to be illogical since at least from 2016 onwards, actions have been taken to deal with climate change (e.g., Paris Agreement). On the other hand, the social pillar consistently has a leading role in all time periods examined in the analysis.

Future goals are to proceed with the analysis of the years using more recent data, which will allow the examination of the effect of the new conditions and challenges that have emerged since 2019 (e.g., pandemic, geopolitical crises, etc.). The further investigation of the materiality of ESG indicators is also important to understand how ESG affects corporate performance and the investment outcomes for investors. Finally, a comparison of the multicriteria methodology presented in this study with other approaches could be performed to examine the quality of the results and the suitability of different analytical tools in this area.

References

Angelopoulos, D., Siskos, Y., & Psarras, J. (2019). Disaggregating time series on multiple criteria for robust forecasting: The case of long-term electricity demand in Greece. *European Journal of Operational Research, 275*(1), 252–265. https://doi.org/10.1016/j.ejor.2018.11.003

Aupperle, K. E., Carroll, A. B., & Hatfield, J. D. (1985). An empirical examination of the relationship between corporate social responsibility and profitability. *Academy of Management Journal, 28*(2), 446–463. https://doi.org/10.2307/256210

Doumpos, M., Andriosopoulos, K., Galariotis, E., Makridou, G., & Zopounidis, C. (2017). Corporate failure prediction in the European energy sector: A multicriteria approach and the effect of country characteristics. *European Journal of Operational Research, 262*(1), 347–360. https://doi.org/10.1016/j.ejor.2017.04.024

Doumpos, M., & Zopounidis, C. (2002). *Multicriteria decision aid classification methods* (Vol. 73). Springer. https://doi.org/10.1007/b101986

Escrig-Olmedo, E., Fernández-Izquierdo, M., Ferrero-Ferrero, I., Rivera-Lirio, J., & Muñoz-Torres, M. (2019). Rating the raters: Evaluating how ESG rating agencies integrate sustainability principles. *Sustainability, 11*(3), 915. https://doi.org/10.3390/su11030915

Giese, G., Lee, L.E., Melas, D., Nagy, Z., & Nishikawa, L. (2019). Foundations of ESG investing: How ESG affects equity valuation, risk, and performance. *The Journal of Portfolio Management, 45*(5), 69–83. https://doi.org/10.3905/jpm.2019.45.5.069

Jacquet-Lagreze, E., & Siskos, J. (1982). Assessing a set of additive utility functions for multicriteria decision-making, the UTA method. *European Journal of Operational Research, 10*(2), 151–164. https://doi.org/10.1016/0377-2217(82)90155-2

Jacquet-Lagrèze, E., & Siskos, Y. (2001). Preference disaggregation: 20 years of MCDA experience. *European Journal of Operational Research, 130*(2), 233–245. https://doi.org/10.1016/S0377-2217(00)00035-7

Khaled, R., Ali, H., & Mohamed, E. K. A. (2021). The sustainable development goals and corporate sustainability performance: Mapping, extent and determinants. *Journal of Cleaner Production, 311*, 127599. https://doi.org/10.1016/j.jclepro.2021.127599

Muñoz-Torres, M. J., Fernández-Izquierdo, M. Á., Rivera-Lirio, J. M., & Escrig-Olmedo, E. (2019). Can environmental, social, and governance rating agencies favor business models that promote a more sustainable development? *Corporate Social Responsibility and Environmental Management, 26*(2), 439–452. https://doi.org/10.1002/csr.1695

Razzaq, A., Ajaz, T., Li, J. C., Irfan, M., & Suksatan, W. (2021). Investigating the asymmetric linkages between infrastructure development, green innovation, and consumption-based material footprint: Novel empirical estimations from highly resource-consuming economies. *Resources Policy, 74*, 102302. https://doi.org/10.1016/j.resourpol.2021.102302

Siskos, Y., Grigoroudis, E., & Matsatsinis, N. F. (2005). UTA methods. In *Multiple criteria decision analysis: State of the art surveys* (pp. 297–334). Springer. https://doi.org/10.1007/0-387-23081-5_8

Syed, A. M. (2017). Environment, social, and governance (ESG) criteria and preference of managers. *Cogent Business & Management, 4*(1), 1340820. https://doi.org/10.1080/23311975.2017.1340820

Tikhonov, A. N., Goncharsky, A. V., Stepanov, V. V., & Yagola, A. G. (1995). *Numerical methods for the solution of Ill-posed problems*. Springer Netherlands. https://doi.org/10.1007/978-94-015-8480-7

Yang, Q., Du, Q., Razzaq, A., & Shang, Y. (2022). How volatility in green financing, clean energy, and green economic practices derive sustainable performance through ESG indicators? A sectoral study of G7 countries. *Resources Policy, 75*, 102526. https://doi.org/10.1016/j.resourpol.2021.102526

Zhao, C., Guo, Y., Yuan, J., Wu, M., Li, D., Zhou, Y., & Kang, J. (2018). ESG and corporate financial performance: Empirical evidence from China's listed power generation companies. *Sustainability, 10*(8), 2607. https://doi.org/10.3390/su10082607

Lufthansa Airlines. The Microeconomic and Macroeconomic Environment of the Company and the Industry in 2020 and Its Readiness Against Crisis

Spyros Zervas

Abstract The objective of this work is to provide an analysis of the impact of the microeconomic and macroeconomic environment of Lufthansa Airlines and sets out the performance of the organization in relation to its main competitors from 2006 until 2019. The airlines industry represents an oligopoly market with high barriers to enter and consists of large commercial aviation companies, top ten of them accounting for approximately half of the global business. For the assessment of the oligopoly market of the airlines industry, respective economic theory, and models are used (Besanko in Economics of strategy. Wiley, Hoboken, 2015; Hall in Price theory and business behavior, pp. 12–45, 1939). To explore the company's market exposures and its cost vulnerabilities, a dedicated analysis of the top ten airlines in the world, is taking place and is analyzed by the application of various economic models (Eiteman in Am Econ Rev 42(5): 832–838, 1952; Samuelson in Microeconomics. McGraw-Hill, p. 110, 2001; Begg in Economics for Business. Macgrew Hill, Maidenhead, 2020). The demand of airlines is exposed to various elasticities depending on the region and the type of the airline while the evolution of oil prices contains one of the most important factors of industry and company's profitability. Therefore, a detailed analysis of the macroeconomic impact of oil prices is performed. Lufthansa has a small sales margin due to the high costs and is a company relatively more vulnerable and exposed to external economic shocks due to its structural business design. Because of that it is not well equipped to withstand global economic shocks. Lufthansa needs to transform its business model in order to survive future crises by optimizing its operating costs and further growing its scale.

Keywords Macroeconomics · Lufthansa · Airlines · Economics · Airlines oligopoly · Airlines competition · Economic shocks · Cost vulnerability

---

S. Zervas (✉)
Warwick Business School, The University of Warwick, Coventry, UK
e-mail: spyros.zervas@warwick.ac.uk

C. Zopounidis et al. (eds.), *Operational Research Methods in Business, Finance and Economics*, Lecture Notes in Operations Research,
https://doi.org/10.1007/978-3-031-31241-0_8

1 Methodologies, Data and Analysis

Data from secondary sources such as the annual reports of Lufthansa from 2006 to 2020 for the examined period were collected. Moreover, respective literature and industry reports, and information in Airlines company websites were assessed and examined. Additionally, we analysed the qualitative and quantitative data collected from all the other major commercial aviation companies.

Among several models, we decided to use the economic frameworks that contain the gravitas to be leveraged in the practical analysis needed of the competitive nature of the airlines industry.

For example, the market of the airlines industry is assessed, by using the market oligopoly framework (Besanko, 2015) that is also explored in further detail by applying the kinked demand curve model (Hall & Hitch, 1939; Stigler, 1947; Sweezy, 1939) in more dynamic situations.

To explore the company's macroeconomic exposure, an analysis of past data of oil prices is performed (OECD, 2019), while for the market exposures and its cost vulnerabilities, a dedicated report of the top ten commercial airlines in the world, is formed and analyzed accordingly through the lens of economic models of cost mechanics such as fixed and variable costs and return labor costs analysis (Eiteman & Guthrie, 1952; Samuelson & Nordhaus, 2001; Begg & Ward, 2020). In order to assess the dynamics of the demand that the airlines industry is exposed to, a dedicated secondary elasticities analysis report for IATA is used to examine the various elasticities depending on the geographies and the types of the airlines (Inter VISTAS Consulting, 2007).

2 Introduction

Lufthansa AG is one of the biggest aviation groups worldwide, the largest German airline and together with its network airlines, the second largest airline in Europe. Lufthansa operates in all four categories of airline industry (international, national, regional and cargo) and is also one of the founding members of "Star Alliance", world's biggest airline alliance, founded in 1997 (Lufthansa, 2020).

Lufthansa Group, besides Lufthansa Airlines, also owns (i) subsidiary passenger airlines (Austrian Airlines, Swiss Air, Brussels Airlines, and Euro-wings and (ii) other aviation-related companies, such as Lufthansa Technik and LSG Sky Chefs with a fleet of 763 aircrafts in total. Lufthansa corporate headquarters are in Cologne while the company also owns a couple of operation hubs at Frankfurt and Munich Airports (Lufthansa, 2020).

The airline used to be a state-owned enterprise and the official German airline until 1994. Looking at its current shareholder synthesis (data from the end of 2019), German investors hold 67.3% of the shares followed by shareholders from Luxembourg (10.4%) and investors from the US (8.1%).

2.1 Lufthansa Overall Performance (2006–2019)

LH group experienced sustainable top-line growth from 2006 to 2019, growing 84% in total, during those 14 years. Sales growth was driven by a combination of acquiring new regional airlines in Europe (Austrian, Swiss Air etc.) & organic growth of Lufthansa core business as a result of increasing their fleet and expanding the number of international flights' itineraries. In terms of profit/loss, LH delivers on average a profit of 3–4% of sales, reaching €1.21 Bn of profit in 2019 (Table 1; Fig. 1).

Given that the company's profitability is highly correlated to the oil prices, during the period between 2011 and 2014 (oil crisis—OECD, 2019), Lufthansa experienced the lowest profits.

3 Airlines Market Exposure and Nature of Competition

3.1 Airlines Oligopoly

3.1.1 Overall Macroeconomic Assessment of Demand and Supply Determinants of Airline Industry

The airline industry is characterized by oligopoly. In this market context, a relatively small number of airlines provide similar services to consumers. Moreover, the industry is exposed to various determinants that may affect the demand and supply. The main determinants on demand curve are (i) consumer income, (ii) prices of related goods like fuel, (iii) seasonality, (iv) total number of buyers, (v) expectation on future income and evolution of future prices, and others. The oligopoly of the airlines' industry is partially related to the high barriers for new companies to enter and the low profit margin. Therefore, Airlines have very low return of equity (ROE) and return of capital employed (ROCE) (Table 2).

3.1.2 Oligopoly and Barriers to Entry

The most important barrier to enter is the high fixed cost, such as fuel and advertising costs, as well as costs related to the airline staff (management, pilots, hostesses etc.). This has as a result a very low profit margin which requires high scale and top-line growth to generate a sustainable profit stream. In order to reach this high scale, airlines may need to build a very strong network that would require a large capital investment (Fig. 2).

The kinked demand curve model (Hall & Hitch, 1939), can be applied in this oligopolistic market of commercial aviation in order to illustrate the "stickiness" of prices. In our case, if Lufthansa increases the fair of the tickets to one of their flights,

Table 1 Lufthansa group annual sales and profit 2006–2019 (Lufthansa, 2006–2019)

	2006	2007	2008	2009	2010	2011	2012	2013	2014	2015	2016	2017	2018	2019
Sales[a]	19.8	22.4	24.9	22.3	27.3	28.7	30.1	30.0	30.0	32.1	31.7	35.6	35.6	36.4
Net P/L[a]	0.80	1.65	0.54	−0.04	1.13	−0.01	0.99	0.31	0.06	1.69	1.77	2.36	2.16	1.21
No.[c] of employees[b]	94.5	105.3	107.8	117.5	117.0	116.4	117.0	118.3	118.8	120.7	124.3	129.4	135.5	138.4
No.[c] of passengers[b]	53.4	62.9	70.5	77.3	91.2	100.5	103.1	104.6	106.0	107.7	109.7	130.0	141.9	145.1
No. of aircrafts[b]	430	513	534	722	710	636	627	622	615	602	617	728	763	763

[a]Figures in Billion Euros (Bn€)
[b]At the end of the year
[c]Figures in Thousands

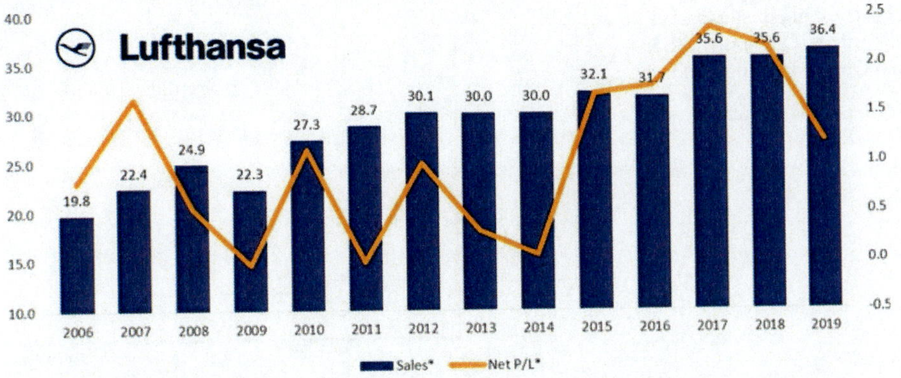

Fig. 1 Lufthansa group annual sales and profit evolution 2006–2019. *Figures in Billion Euros (Bn€) (Lufthansa, 2006–2019)

Table 2 Levels for ROE (Return on Equity), ROCE (Return on capital employed), Net Margin and ATO (Asset Turnover) (Standard & Poor's, 2020)

	ROE (%)	ROCE (%)	Net Margin (%)	ATO
Restaurants	15.6	14.2	5.0	2.83
Printing and publishing	14.6	14.3	6.5	2.20
Business services	14.6	15.3	5.2	2.95
Chemicals	14.3	13.6	7.1	1.91
Food stores	13.8	12.6	1.7	7.39
Road Transport	13.8	10.9	3.8	2.88
Food products	13.7	12.1	4.4	2.74
Communications	13.4	9.5	12.5	0.76
General Stores	13.2	12.4	3.5	3.55
Petroleum refining	12.6	11.8	6.0	1.96
Transportation equipment	12.5	11.1	4.5	2.47
Airlines	12.4	8.6	4.3	1.99
Utilities	12.4	8.5	14.5	0.59
Wholesalers, non-durable goods	12.2	8.6	2.3	3.72

direct competitors operating in the same region may leave their prices the same, as they hope to increase their market share by capturing customers from LH (Fig. 3). On the other hand, if LH decreases the price for the same flight, competition, may follow as they would try to avoid losing market share. Therefore, the demand curve when prices are cut is relatively price inelastic. Looking at the kinked demand curve model we can observe that prices in an oligopolistic market, such as the airline industry, are likely to be relatively "sticky". If marginal costs increase for example from MC1 to

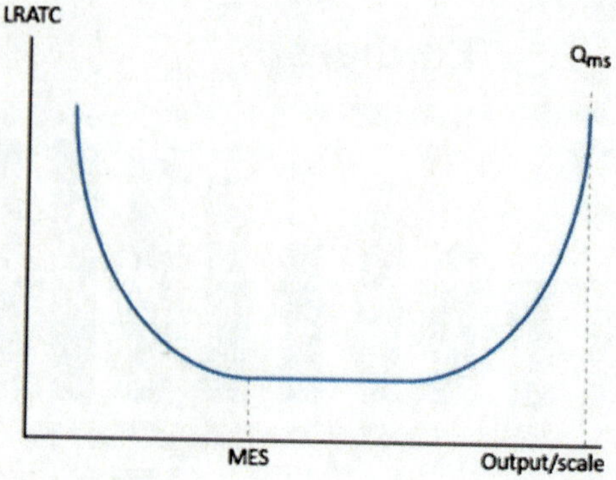

Fig. 2 Oligopolistic market of Airlines. Oligopoly, MES and Qms (Besanko, 2015)

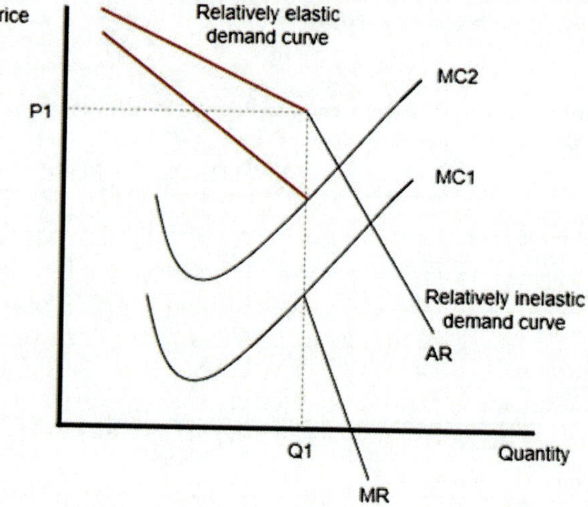

Fig. 3 The kinked demand curve model applied oligopolistic market of Airlines (Hall & Hitch1939; Stigler, 1947)

MC2, the price will stay the same. Therefore, LH is likely to absorb the costs and reduce the profit (Fig. 3).

3.2 *Performance Versus Competition*

There are more than 5000 national, international, domestic, or other airlines in the world. Nevertheless, only 10 of the biggest commercial aviation companies make ~ 45% of the total annual revenue that is calculated to $ 812 Bn (Statista, 2018; Table 3).

LH Group is the 4th biggest airline in the world in absolute sales (2018), and within the 6 largest airlines in terms of profit. (Table 3). The top 10 airlines include: Delta, South-West, FEDEX, United and American Airlines based in the US, China South Airlines and Emirates based in Asia and the remaining three in Europe. In

Table 3 Top 10 most profitable Airlines in the world

Airline	Annual sales	Rank in sales	Annual profit	Profit margin %*
DELTA Airlines	$44.4 Bn	3	$5.2 Bn	11.7
Southwest Airlines	$21.9 Bn	9	$2.5 Bn	11.2
International Airlines Group	$28.3 Bn	8	$2.9 Bn	10.2
FED EX	$65.5 Bn	1	$4.6 Bn	6.9
Lufthansa Group	$41.3 Bn	4	$2.5 Bn	6.1
UNITED Airlines	$37.7 Bn	5	$2.1 Bn	5.6
American Airlines	$44.6 Bn	2	$1.9 Bn	4.3
China South Airlines	$21.7 Bn	10	$0.5 Bn	2.3
Air France – KLM	$30.7 Bn	6	$0.6 Bn	1.5
Emirates	$29.8 Bn	7	$0.2 Bn	0.8

*Profit Margin Calculated: gross annual profit/gross annual sales
(Lufthansa Group 2019; Air France—KLM, 2019; United Airlines, 2019; Southwest, 2019; IAG, 2019; Delta Airlines, 2019; FEDEX, 2019; American Airlines, 2019; China Airlines, 2019; Emirates, 2019)

Europe all three big aviation companies, Lufthansa group, International Airlines Group (IAG) and Air France—KLM Group are network airlines' groups that consist of multiple, sizeable individual airlines. Specifically, Lufthansa Group owns together with Lufthansa, Austrian, Swiss and Brussels and others, IAG owns British Airways, Iberia, Aer Lingus, Vueling and others and Air France-KLM Group owns those two airlines.

According to a report from the Organization for Economic Co-operation and Development in 2014 regarding Airlines Competition, there are two key areas of competition, quality, and price. With the founding of EasyJet in the mid-90s and the expansion of other low-cost airlines in Europe like Ryanair, WizzAir and others, price became a very important element. This led to higher demand price elasticity and more airlines merged or created alliances and networks to strengthen their position against this trend. Moreover, as part of those alliances, most of the big commercial airlines developed loyalty programs to reward frequent consumers with complementary benefits (access in lounges, boarding priority etc.)

3.3 *Product Market Exposure and Demand Price Elasticity*

Looking at the market exposure of Lufthansa we are going to assess the elasticity of its pricing. There is a debate in economics if the demand of this industry is elastic or inelastic. The answer we give is that it depends on the region, the type, and the competition of the airline.

The demand of airline tickets may be inelastic overall, as explained also in the kinked demand model above, given that traveling with an airplane is a specific need that traditionally only a few, similar companies could serve. However, given that nowadays there are multiple choices from national, international, and domestic airlines with a variety of fairs (having also low-cost options) and quality of service, that are operating the same itineraries, consumers may be more sensitive to pricing.

An important variable is the area scope of the airline, i.e. the region that each airline is covering. According to a report by Inter VISTAS Consulting prepared for IATA (Table 4), we see that areas like intra sub-Sahara and Intra South America have inelastic demands due to limited flights in those regions, while in regions like Europe the price elasticity is high (Inter VISTAS Consulting, 2007). All Lufthansa Group airlines have their hubs and as a result their main operations, in Europe. Given that there is a high competition and a plethora of flights from airlines to cover the increased consumer demand, Lufthansa is exposed to a relatively higher price elasticity (Fig. 4).

Another characteristic of the pricing of the airlines is that they are exposed to dynamic pricing (Fig. 5). Specifically, the airlines (i) adjust the pricing real-time based on seasons and consumer demand and (ii) may add/change price tiers (premium and discount offers) to maximize the sales even last minute.

Table 4 Aviation demand elasticities. Estimating Air Travel demand elasticities final report. Prepared for IATA by Inter VISTAS Consulting (Inter VISTAS Consulting, 2007)

	Route/market level		National level		Pan-National level	
	Short-haul	Long-haul	Short-haul	Long-haul	Short-haul	Long-haul
Intra North America	−1.54	−1.40	−0.88	−0.80	−0.66	−0.60
Intra Europe	−1.96[a]	−1.96	−1.23	−1.12	−0.92	−0.84
Intra Asia	−1.46	−1.33	−0.84	−0.76	−0.63	−0.57
Intra Sub-Sahara Africa	−0.92	−0.84	−0.53	−0.48	−0.40	−0.36
Intra South America	−1.93	−1.75	−1.10	−1.00	−0.83	−0.75
Trans Atlantic (North America–Europe)	−1.85	−1.68	−1.06	−0.96	−0.79	−0.72
Trans Pacific (North America–Asia)	−0.92	−0.84	−0.53	−0.48	−0.40	−0.36
Europe–Asia	−1.39	−1.26	−0.79	−0.72	−0.59	−0.54

[a]The short-haul adjustor has not been applied to the Intra Europe short-haul elasticity in order to maintain elasticities below 2.0

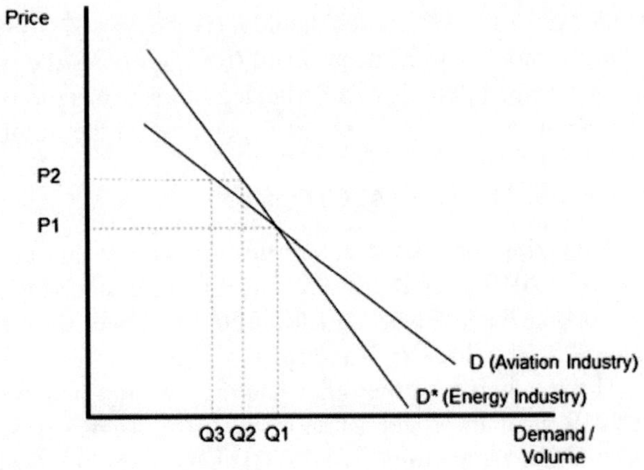

Fig. 4 Price/volume demand curve of Aviation industry versus Energy industry

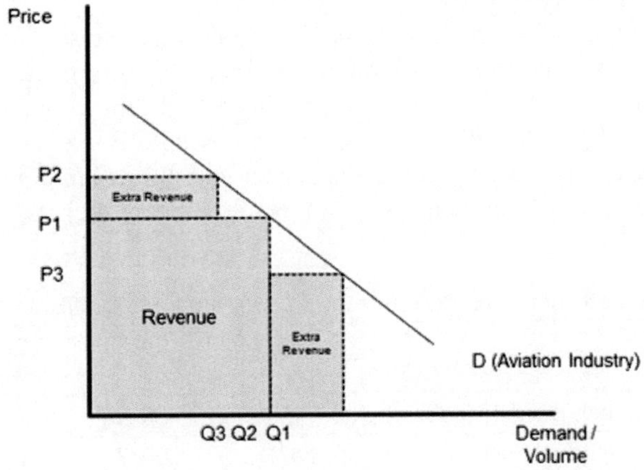

Fig. 5 Revenue generation from dynamic pricing in Aviation industry

4 Vulnerability and Costs

4.1 Cost Structure

In 2019, Lufthansa reported the total amount of €37 Bn in operating expenses, 24.5% of which being labour cost followed by cost of fuel (18%) and cost of other materials and services (summing to 35%).

- 53% of the total cost is covering materials and expenses like fuel, fees and charges related to the core product of Lufthansa, flights. Most of this cost is variable as it depends on the number of flights; the bigger the amount of flights (leading to higher sales), the higher the variable cost (VC) leading to an average variable cost if divided by sales (AVC).

- The rest of the expenses include: (i) staff costs, (ii) depreciation of assets (buildings, fleet of aircrafts etc.) and (ii) other costs like maintenance. Assuming that the company employs a dedicated number of employees in the short-run and its paying salary-scheme is not depending on the number of flights, all those costs are considered to be fixed (FC) in the short run, leading to an average fixed cost if divided by sales (AFC) (Table 5).

Analyzing the cost structure further, we see that the total variable cost grew by 8% from 2019 to 2018, as a result of getting more flights of Lufthansa or its group network airlines (Swiss, Austrian etc.) to cover the increased demand (growth in sales, ~ 2.3%) (Figs. 6, 7 and 8).

The fixed cost on the other hand, grew significantly less (by 3%) and the AFC in 2019 was almost the same as in 2018, at 44% (Fig. 13). Given that we see an increasing average total cost (ATC) as the sales grew, we can conclude that the total cost of Lufthansa expenses is positioned at the decreasing labor return (DLR) of the cost curve (Fig. 9).

One of the main reasons that the cost structure is dynamic is the phenomenon of "stepped fixed cost" related to the capacity of the aircrafts. For LH to maximize the income generated by every single flight for example, they need to always have the flights as full of passengers, as possible. If we get a scenario that all the flights of Lufthansa have the same capacity and they are always full and there is still an underserved number of customers looking for tickets, LH may potentially acquire

Table 5 Lufthansa Cost structure 2019 versus 2018 (Lufthansa, 2019)

	2019 in € m	2018 in € m	Change in %	Share of operating expenses in%
Cost of materials and services	19,827	18,367	8	53
of which fuel	6715	6087	10	18
of which fees and charges	4523	4457	1	12
of which external services MRO	1911	1848	3	5
of which charter expenses[a]	814	718	13	2
Staff costs[b]	9111	8924	2	25
Depreciation[c]	2692	2180	23	7
Other operating expenses[d]	5494	5693	−3	15
of which indirect staff costs and external staff	1201	1226	−2	3
of which rental and maintenance expenses	742	923	−20	2
Total operating expenses	*37,124*	*35,164*	6	*100*

[a]In 2018 including operating lease expenses according to IAS 17
[b]Without past service costs/settlement
[c]Without impairment losses
[d]Without book losses and write-downs on assets held for sale

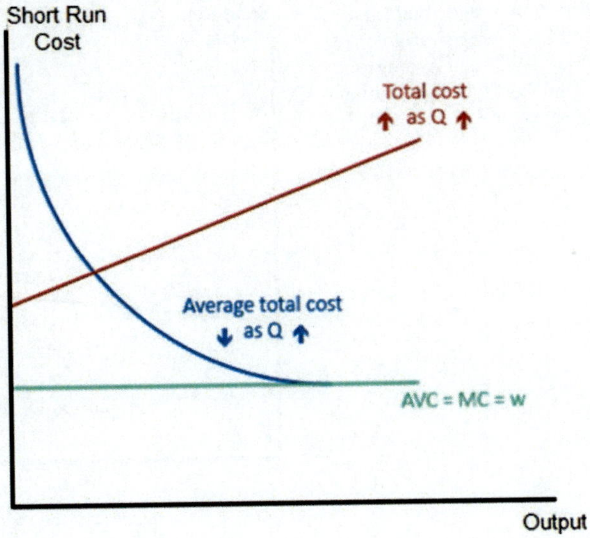

Fig. 6 Example basic cost mechanics

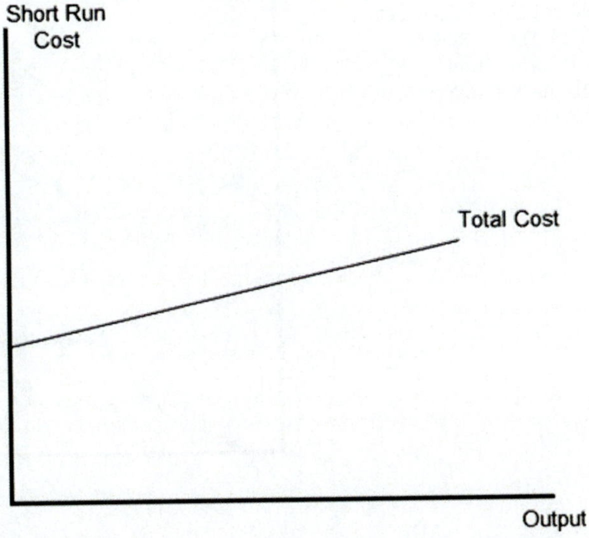

Fig. 7 Total cost evolution of Lufthansa Group 2019 versus 2018 (Lufthansa, 2019)

additional aircrafts to cover this extra demand. Until they'd manage to maximize the capacity of the new flights operated by the new aircrafts this would lead to an increase of the cost per flight and ATC. Given that in 2018 LH grew their fleet by 35 aircrafts, this justifies the decreasing labor return (DLR).

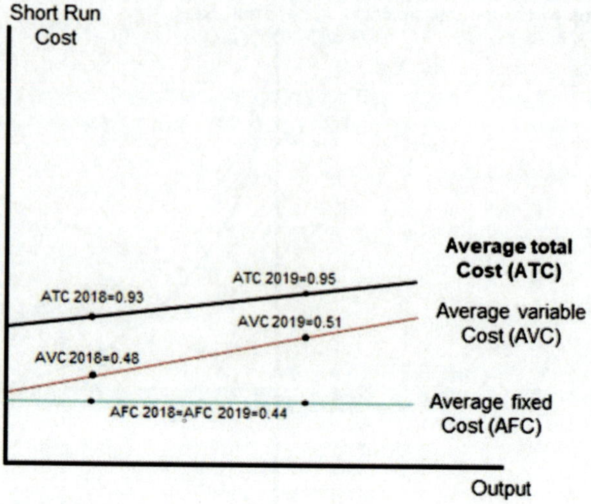

Fig. 8 Average Total Cost (ATC), Average Variable Cost (AVC) and Average Fixed Cost (AFC) evolution in short run (2019 vs. 2018)

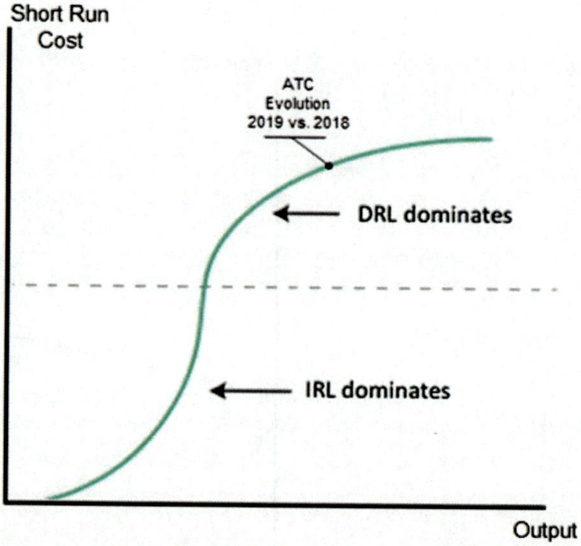

Fig. 9 Return labor matrix. Lufthansa has Decreasing Return Labor (DRL). Data: 2019 Annual Report Lufthansa Group (Samuelson & Nordhaus, 2001)

4.2 Cost Vulnerabilities

The airlines industry generally experiences high dependance on oil/fuel, as well as high fixed and quasi-fixed costs related with the aircrafts, maintenance, and variable costs (with labor cost being the most significant one). Therefore, the industry is exposed to both high Type 2 (by nature) and Type 1 vulnerabilities. Moreover, the high operational costs lead to limited profit margin, so in order for the airline business model to make financial sense, it requires high topline volumes.

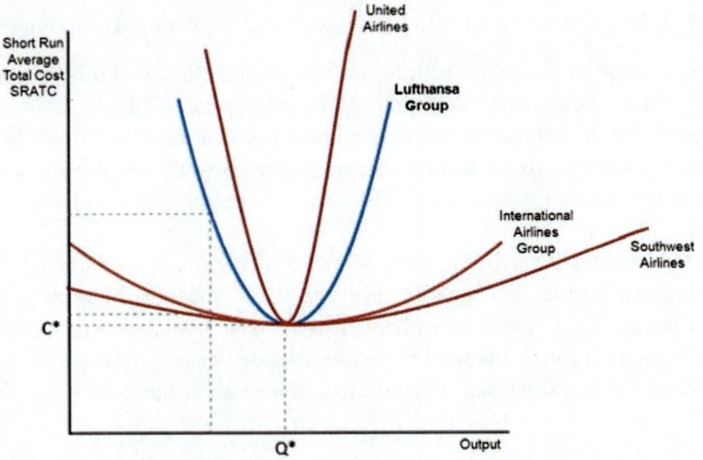

Fig. 10 Business vulnerability and the slope of the SRATC curve (Eiteman & Guthrie, 1952)

After consolidating data from top 10 airlines in the world in absolute profit figures (Table 3), we see that Lufthansa, although was number 4 in absolute sales (2018), is 5th in terms of profit margin %.

If we get a closer look to the three companies that together with Lufthansa, deliver similar profit figures, i.e. between \$2.1 and \$2.9 Bn (Tables 1 and 3), we see that Lufthansa experiences the highest annual gross sales but is third in terms of sales margin (Table 3).

If we assume that those 4 companies have the same minimum cost level of output (Q^*), Lufthansa Group is the second most vulnerable business in terms of Average total cost curve (SRATC) (Fig. 10). With this analysis, we confirm that Lufthansa Group, follows similar patterns with other airlines in terms of costs where the high variable and fixed costs don't allow high profit margin. Within this group of peer companies, LH seems that has relatively higher costs and as a result contains a—ceteris paribus—relatively more vulnerable business to exposure to external pressure or crisis.

5 The Macroeconomic Impact of Oil Prices

5.1 Evolution of Oil Price

The price of oil is an essential element of the Lufthansa cost and all airlines. Only in 2019, 18% of Lufthansa operating expenses was fuel of the aircrafts (Lufthansa, 2019). Therefore, the evolution of oil prices contains an important factor of company's profitability.

Looking at the price levels of oil in the past 14 years, we draw two conclusions.

1. There is a negative correlation between the increase of oil prices and the profitability of Lufthansa. Specifically, the period between 2011 and 2014 when the oil prices were the highest levels in the past 40 years (OECD, 2019), Lufthansa reported the lowest profits, i.e. on yearly average ~ €0.3 Bn, 80% lower versus what it would deliver a year later in 2015 when the oil crisis was over (Figs. 11 and 12).
2. Out of the biggest 10 airlines in the word, 5 of them are based in USA (Table 3). At the same time, US buys on average 9.2% cheaper oil versus Germany (past 14 years, Fig. 11) and 14% cheaper in the past 5 years (OECD, 2019). This generates an additional financial pressure for Lufthansa versus key competitors like United, Delta, American Airlines, Southwest and others.

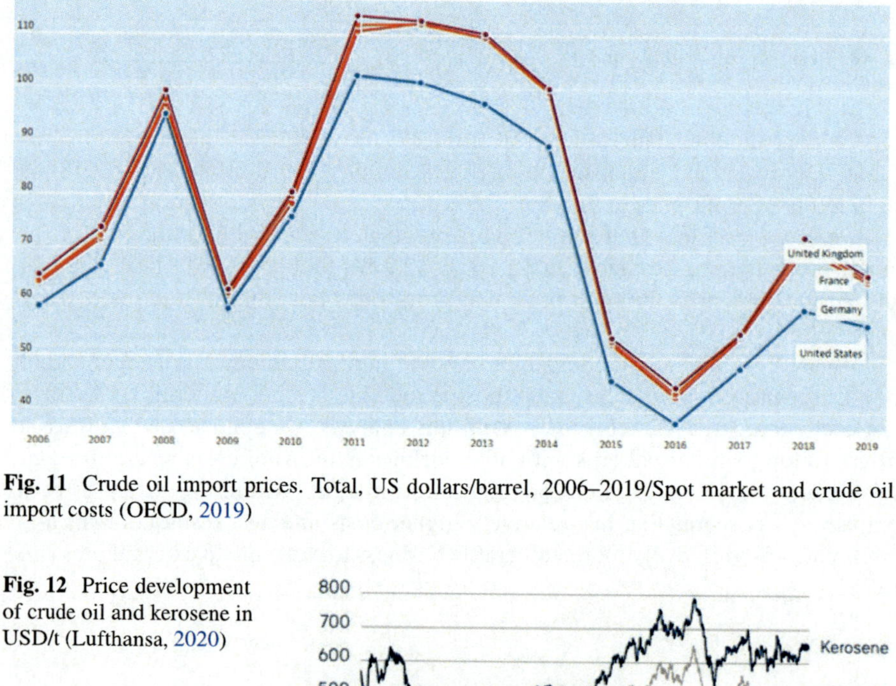

Fig. 11 Crude oil import prices. Total, US dollars/barrel, 2006–2019/Spot market and crude oil import costs (OECD, 2019)

Fig. 12 Price development of crude oil and kerosene in USD/t (Lufthansa, 2020)

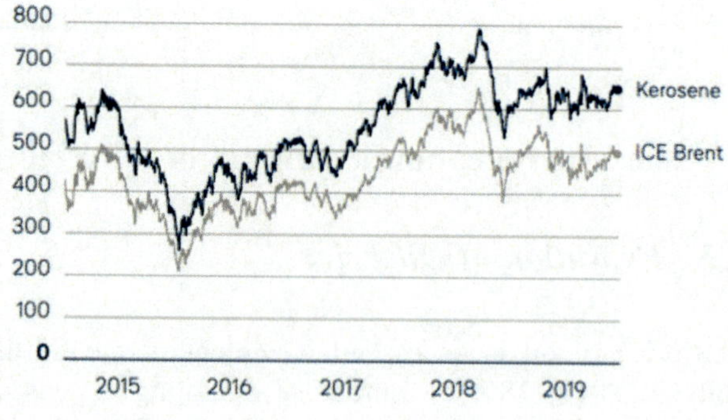

6 Conclusions. Effectiveness of Lufthansa Strategy and Readiness Towards Crisis

6.1 Overall Effectiveness of Business Strategy (2006–2019)

Lufthansa is a large company with small profitability. Their strategy is to keep growing organically and by acquisitions, in order to benefit from the scale that will be created, allowing them to potentially increase their prices and profit margin. This business model contains a volatility risk for a relatively small return on investment and is driven by the following variables:

1. Lufthansa doesn't seem to maximize the flights' supply / demand balance and sequentially the profit of all the flights. Therefore, it is following a business strategy where the revenue stream is pressured by the limited profit margin. If we also add the uncertainty and instability of external factors like the weather and most importantly, oil prices, it's hard for LH to project and deliver high and sustainable profitability.
2. Moreover, Lufthansa cost is positioned at the decreasing labour return (DLR) of the cost curve (Fig. 9) as a result of the fact that more sales, increase the average total cost (ATC), at least in the short run. This creates a cycle where more sales may lead to buying more aircrafts, which until their capacity is maximized, they drive the average cost up.

Net, considering the above and the overall cost vulnerability of the company explained above, Lufthansa business strategy is not as effective compared to their peers, i.e., airlines with similar size due to higher costs, i.e., higher exposure to oil prices and higher operating costs.

6.2 Lufthansa Readiness Against Economic Shocks (2009, 2020)

2009 was one of the worst financial years for all airlines and LH Group. Although the company managed to deliver €22.3 Bn in top-line, depicted losses in the bottom line (Table 6).

According to an IATA analysis published to centre for aviation, as a result of the global financial crisis, total aviation industry lost 2 years of growth. In the same report, we see that although the demand for flights started showing signs of improvement at the end of the year, there was higher price sensitivity and elasticity forcing large carriers to decrease their prices in order to maintain some of their previous (loyal) customers (IATA, 2009).

From the analysis above comparing Lufthansa with some airlines of similar size, we concluded that the company is relatively more vulnerable and exposed to external economic shocks by its structural business model design. Moreover, as a European

Table 6 Lufthansa group annual sales and profit evolution 2006–2012 (Lufthansa, 2020)

	2006	2007	2008	2009	2010	2011	2012
Sales*	19.8	22.4	24.9	22.3	27.3	28.7	30.1
Net P/L*	0.80	1.65	0.54	−0.04	1.13	−0.01	0.99

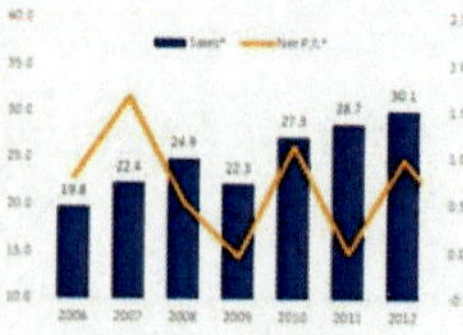

*Figures in Billion Euros (Bn€)

airline, LH is relatively more exposed to turbulence of oil prices versus the American based airlines.

Lufthansa financial weaknesses were exposed in the financial crisis of 2009; the company didn't manage to compete versus low-cost airlines and the drop in their prices reduced the profit margin and led to losses in the P/L reports. However, LH did some pivotal changes; Right after the big shock, the company announced a series of cost cuts of €1.4 Bn (Reuters, 2009), while during the same year it completed the purchase of three airlines in Europe (Swiss, Austrian and Brussels, 2007–2009). This gave the company the opportunity to grow rapidly its scale, redesign its business model and all this with smaller competition as it acquired some of the companies that it was previously competing with for flights in central Europe.

In 2020, a similar situation is observed due to COVID pandemic. The German government offered to LH a €9 Bn bailout to support the airline through this economic crisis (Lufthansa, 2020). Nevertheless, Lufthansa still needs to do some deep changes and transform itself. Those changes should include:

1. The optimization of the operating costs of the company (staff, fuel, loyalty program etc.) and
2. Further growth of its overall footprint in Europe by (i) acquiring smaller airlines in countries where it doesn't have direct access and (ii) increasing the diversification of its portfolio, i.e., Cargo (that during COVID remains a solid revenue stream), low-cost proposition (Eurowings) and further improvement of the quality (and the prices) of their premium customers.

Appendix 1

Calculation of adjusted ROCE and cost of Capital. 2019 versus 2018.

Source: 2019 Annual Report Lufthansa Group, lufthansagroup.com.

in €m	2019	2018	Change in %
Revenue	36,424	35,542	2

(continued)

(continued)

in €m	2019	2018	Change in %
Other operating income	2574	2349	10
Operating income	*38,998*	*37,891*	*3*
Operating expenses	37,309	35,091	6
Result from equity investments	168	174	−3
EBIT	*1857*	*2974*	*−38*
Adjusted EBIT	*2026*	*2836*	*−29*
Interest on liquidity	79	68	16
Taxes (assumption 25% of EBIT + Interest on liquidity)	−484	−761	36
Cost of capital[a]	−1007	−860	−17
EACC	*445*	*1422*	*−69*
ROCE[b] in %	*6.1*	*11.1*	*−5.0 pts*
Adjusted ROCE[c] in %	*6.6*	*10.6*	*−4.0 pts*
Balance sheet total	42,659	38,213	12
Non-interest bearing liabilities			
of which liabilities from unused flight documents	4071	3969	3
of which trade payables, other financial liabilities, other provisions	E.B68	6306	−7
of which advance payments, deferred income, other non-financial liabilities	3089	2830	9
of which others	4575	4099	12
Capital employed	25,056	21,009	19
Average capital employed[d]	*23,982*	*20,502*	*17*
WACC in %	4.2	4.2	–

[a]WACC × Average capital employed
[b](EBIT + Interest on liquidity—25% taxes)/Average capital employed
[c](Adjusted EBIT + Interest on liquidity—25% taxes)/Average capital employed
[d]Average capital employed in 2019 including IFRS 16 right-of-use assets as of 1 January 2019

Appendix 2

Germany Gross domestic product (GDP). Total, Million US dollars, 2006–2019.
Source: Aggregate National Accounts, SNA 2008 (or SNA 1993): Gross domestic product. Retrieved from OECD Data. data.oecd.org.

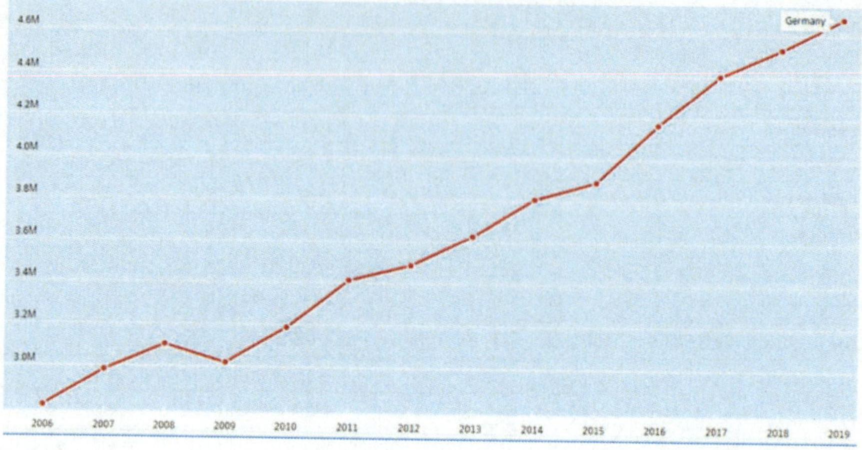

Appendix 3

Investment in airport infrastructure in Germany 2004–2018 Published by Statista Research Department, May 25, 2020.

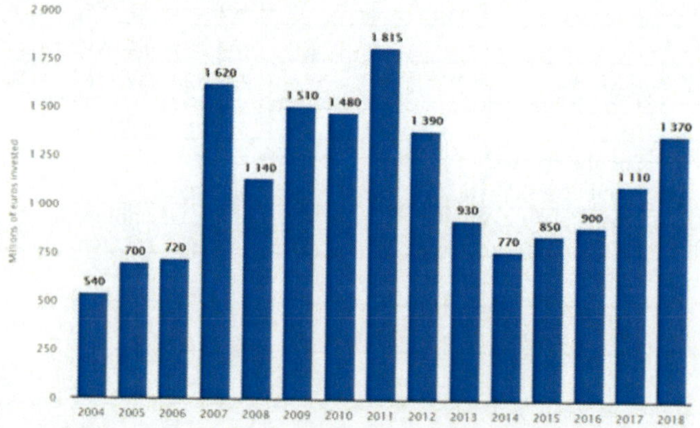

This statistic illustrates the total amount invested in airport infrastructure in Germany from 2004 to 2018, in million euros. In the period of consideration, airport infrastructure investments oscillated. In 2018, investments in this sector amounted to over 1.37 billion euros. The largest amount of investments in airport infrastructure was recorded in 2011, at a total of approximately 1.8 billion euros. Amount of money invested in airport infrastructure in Germany from 2004 to 2018 (in million euros).

References

Begg, D., & Ward, D. (2020). *Economics for business* (6th ed.). Macgrew Hill.

Besanko, D., Dranove, D., & Shanley, M. (2015). *Economics of strategy* (7th edn). Wiley. ISBN 9781119042310.

CEIC. (2020). *Germany private consumption: % of GDP 1991–2020*. Quarterly%, CEIC Data.

Center for Aviation. (2009). *Global economic crisis has cost the aviation industry two years of growth: IATA*. centreforaviation.com.

Dempsey, P. S. (2017). *Introduction to airline economics*. McGill University Institute of Air & Space Law.

Eiteman, W. J., & Guthrie, G. E. (1952). The shape of the average cost curve. *American Economic Review, 42*(5), 832–838. JSTOR 1812530.

Europe Intelligence Wire. (2006). *Lufthansa opens new office complex in Frankfurt (Lufthansa eroffnet neue Konzernzentrale in Frankfurt)*. Retrieved August 25, 2009.

Gagey, F. (2018). *AirFrance KLM Group financial review 2018*. Full year 2018 results. AirFrance KLM Group. airfranceklm.com, p. 9.

Gupta, N. (2019). *10 most profitable airlines in the world: 2019 rankings*. Source: 2018 P&Ls. Lists, News. Insider Monkey Blog. insidermonkey.com.

Hall, R., & Hitch, C. (1939). Price theory and business behavior. *Oxford Economic Papers, 2*, 12–45.

Heinz, B., (1994). *Neue Freiheit*. Die Zeit (in German). Archived from the original on 21 October 2013. Retrieved October 21, 2013. Wikipedia.

Inter VISTAS Consulting Inc. (2007). *Estimating air travel demand elasticities final report*. Prepared for IATA by Inter VISTAS Consulting.

International Airlines Group. (2018). *Annual report and accounts 2018*. IAG. iairgroup.com.

Lufthansa. (2011). *We hereby invite our shareholders to attend the 51st Annual General Meeting* (PDF). investor-relations.lufthansa.com. Retrieved August 25, 2009.

Lufthansa (2006). *How to get there*. lac.lufthansa.com. Retrieved July 30, 2002.

Lufthansa. (2020). *Lufthansa Group Company Portrait*. www.lufthansagroup.com.

Lufthansa Group. (2006–2019). *Annual reports 2006–2019*. Lufthansagroup.com. LH Group.

Market Watch. (2018a). *Southwest Airlines Co. Annual Financials for Southwest Airlines Co.* marketwatch.com.

Market Watch. (2018b). *Annual financials for Delta Air Lines Inc*. marketwatch.com.

OECD Data. (2019a). *Crude oil import prices. Total, US dollars/barrel, 2006–2019a*. Source: spot market and crude oil import costs: Crude oil import costs and index. Retrieved from OECD Data. data.oecd.org.

OECD Data. (2019b). *Gross domestic product (GDP) Total, Million US dollars, 2006–2019b*. Source: Aggregate National Accounts, SNA 2008 (or SNA 1993): Gross domestic product. Retrieved from OECD Data. data.oecd.org.

Organization for Economic Co-operation and Development. (2014). *Airline competition report*. Directorate For Financial and Enterprise Affairs Competition Committee. OECD. oecd.org.

Plecher, H. (2020). *Germany: Ratio of government expenditure to gross domestic product (GDP) from 2014 to 2024*. Statista.com. Statista.

Rebbe, K. (2019). *Erleben ist das neue Haben* (DE). Retrieved Nov 6, 2019. Wikipedia.

Samuelson, P. A., & Nordhaus, W. D. (2001). *Microeconomics* (17th edn, p. 110). McGraw-Hill. ISBN 0071180664.

Scholtes, B. (2017). *Lufthansa reaches for global skies*. Deutsche Welle. dw.com.

Sheahan, M. R. (2009). *Lufthansa: Winner or loser of economic crisis?* Reuters.com.

Star Alliance. (2014). *Lufthansa*. Retrieved July 5, 2014 Wikipedia. staralliance.com.

Statista. (2018). *Revenue of commercial airlines worldwide from 2003 to 2021* (in billion U.S. dollars). Statista.com.

Stigler, G. (1947). Kinky oligopoly demand and rigid prices. *Journal of Political Economy, 55*, 432–449.

Sweezy, P. (1939). Demand under conditions of oligopoly. *Journal of Political Economy, 47*, 568–573.

S&P Global. (2020). *Airlines brace for COVID-19 impact.* Special Report. spglobal.com.

United Airlines. (2019). *United Airlines reports full-year and fourth-quarter 2018 performance.* United Airlines. hub.united.com.

Wissenbach, I., & Taylor, E. (2020). *Lufthansa soars after top shareholder backs bailout*. Business News. REUTERS. Reuters.com.

World-Bank. (2018). *Germany trade statistics: Exports, imports, products, tariffs, GDP and related development indicator.* World Integrated Trade Solution. World Bank. wits. worldbank.org.